Laboratory Manual to Accompany
Introductory Circuit Analysis

ELEVENTH EDITION

Robert L. Boylestad
Gabriel Kousourou

PEARSON

Prentice
Hall

Upper Saddle River, New Jersey
Columbus, Ohio

NOTICE TO THE READER

Editor in Chief: Vernon Anthony
Production Editor: Rex Davidson
Design Coordinator: Diane Ernsberger
Editorial Assistant: Lara Dimmick
Cover Designer: Candace Rowley
Cover Art: Getty One
Production Manager: Matt Ottenweller
Senior Marketing Manager: Ben Leonard
Marketing Assistant: Les Roberts
Senior Marketing Coordinator: Liz Farrell

This book was set in Times Roman by Carlisle Publishing Sevices and was printed and bound by Courier/Kendallville, Inc. The cover was printed by Coral Graphic Services, Inc.

Pearson Education Ltd.
Pearson Education Singapore Pte. Ltd.
Pearson Education Canada, Ltd.
Pearson Education—Japan

Pearson Education Australia Pty. Limited
Pearson Education North Asia Ltd.
Pearson Educación de Mexico, S.A. de C.V.
Pearson Education Malaysia Pte. Ltd.

10 9 8 7 6 5 4
ISBN: 0-13-219615-8

To our wives, children, and grandchildren

Preface

With each new edition, our first effort is to develop a few new laboratory experiments for the dc and ac sections to offer a broader range of selection. This time, however, we found that most institutions run no more than 14 of the laboratory experiments in each section and that expanding the number of experiments may simply dilute the content of the current laboratory selection. The result is that most of the laboratory experiments from the 10th edition appear in the 11th edition, although most have some small changes in content.

The most noticeable change is the expanded use of tables throughout the dc and ac sections. The use of more tables was suggested by Professor Jim Fiore of Mohawk Valley Community College. At first we were unsure that the change would be an improvement, but after carefully reading his review of the material, it became obvious that this approach could have a very positive impact on the laboratory session. Calculated and measured data can be easily compared rather than having to shift through pages looking for the results of interest. The entire laboratory experiment appeared more compact, friendly, and easier to follow. The result is a major change in basic appearance. Hopefully, those schools that have used the manual over the years will welcome the change. As always, we look forward to your comments, whether positive or negative.

One other major change is the deletion of the last two experiments of each section that dealt with computer methods. We found it more useful to simply insert computer exercises in a number of experiments to permit a comparison between computer methods and calculated or measured results. Because the computer analysis appears at the end of the experiments, it can simply be ignored if the instructor finds the problems unsuitable for the laboratory session or write-up to follow.

TIME ELEMENT

This manual is typically used in programs with laboratory sessions lasting two or three hours. In either case, most institutions begin each session with a short lecture on the theoretical material to be validated, the equipment to be employed, and any pertinent safety regulations.

Each laboratory has sufficient content for a three-hour laboratory session. Because some sessions are limited to two hours, we made every effort to ensure that most parts of each laboratory exercise were free-standing to permit omissions without damaging the flow and primary objectives of the experimental exercise. Of course, if the student is permitted to finish the laboratory report at home, most of the required data can be obtained in a two-hour period.

If a student has properly prepared for the laboratory session by reading the material beforehand and then paying careful attention to the instructor's comments at the introduction to each exercise, he or she should be able to complete the laboratory and answer the questions in a three-hour period.

SOLUTIONS MANUAL

The solutions manual is simply a completed laboratory manual with the data we obtained. This format was chosen to make it easier for the instructor to quickly check the data obtained during the usually busy laboratory session. The new format eliminates the need to search for the page and reference in a manual containing only a tabulated list of data and solutions. Print and online versions.

To access supplementary materials online, instructors need to request an instructor access code. Go to **www.prenhall.com**, click the **Instructor Resource Center** link, and then click **Register Today** for an instructor access code. Within 48 hours after registering you will receive a confirming e-mail including an instructor access code. Once you have received your code, go to the site and log on for full instructions on downloading the materials you wish to use.

ACKNOWLEDGMENTS

The following individuals have given excellent feedback and reviews of previous editions, for which we express our deep appreciation:

James Fiore, Mohawk Valley Community College
Joe Hackney, Collin County Community College
John Jones, Oklahoma State University—Oklahoma City
Hassan Moghbelli, Purdue University—Calumet
Larry Pivec, Daytona Beach Community College
Nilda Watkins, DeVry University

In addition, we wish to thank Lara Dimmick and Rex Davidson at Prentice Hall for their support, input, and patience during the production cycle of the manual.

Robert L. Boylestad
Gabriel Kousourou

Contents

dc Experiments

ac Experiments

Parts List for the Experiments   499
Appendices

Name: _____

Date: _____

Course and Section: _____

Instructor: _____

Math Review and Calculator Fundamentals dc

MATH REVIEW

Fractions

Powers of 10

Scientific Notation

Engineering Notation

Conversions Between Levels of Powers of Ten

Prefixes

Square and Cube Roots

Exponential Functions

Algebraic Manipulations

CALCULATOR FUNDAMENTALS

Initial Settings

Order of Operations

Powers of 10

Exponential Functions

MATH REVIEW

The analysis of dc circuits requires that a number of fundamental mathematical operations be performed on an ongoing basis. Most of the operations are typically covered in a secondary-level education; a few others will be introduced in the early sessions of the lecture portion of this course.

The material to follow is a brief review of the important mathematical procedures along with a few examples and exercises. Derivations and detailed explanations are not included but are left as topics for the course syllabus or as research exercises for the student.

It is strongly suggested that the math review exercises be performed with and without a calculator. Although the modern-day calculator is a marvelous development, the student should understand how to perform the operations in the longhand fashion. Any student of a technical area should also develop a high level of expertise with the use of the calculator through frequent application with a variety of operations. The use of the calculator is covered later in this laboratory exercise.

Fractions

Addition and Subtraction:

Addition and subtraction of fractions require that each term have a common denominator. A direct method of determining the common denominator is simply to multiply the numerator and denominator of each fraction by the denominator of the other fractions. The method of "selecting the least common denominator" will be left for a class lecture or research exercise.

EXAMPLES

$$\frac{1}{2} + \frac{2}{3} = \left(\frac{3}{3}\right)\left(\frac{1}{2}\right) + \left(\frac{2}{2}\right)\left(\frac{2}{3}\right) = \frac{3}{6} + \frac{4}{6} = \frac{7}{6}$$

$$\frac{3}{4} - \frac{2}{5} = \left(\frac{5}{5}\right)\left(\frac{3}{4}\right) - \left(\frac{4}{4}\right)\left(\frac{2}{5}\right) = \frac{15}{20} - \frac{8}{20} = \frac{7}{20}$$

$$\frac{5}{6} + \frac{1}{2} - \frac{3}{4} = \left(\frac{2}{2}\right)\left(\frac{4}{4}\right)\left(\frac{5}{6}\right) + \left(\frac{6}{6}\right)\left(\frac{4}{4}\right)\left(\frac{1}{2}\right) - \left(\frac{6}{6}\right)\left(\frac{2}{2}\right)\left(\frac{3}{4}\right)$$

$$= \frac{40}{48} + \frac{24}{48} - \frac{36}{48} = \frac{28}{48} = \frac{7}{12}$$

(12 being the least common denominator for the three fractions)

Multiplication:

The multiplication of two fractions is straightforward, with the product of the numerator and denominator formed separately, as shown here.

EXAMPLES

$$\left(\frac{1}{3}\right)\left(\frac{4}{5}\right) = \frac{(1)(4)}{(3)(5)} = \frac{4}{15}$$

$$\left(\frac{3}{7}\right)\left(\frac{1}{2}\right)\left(-\frac{4}{5}\right) = \frac{(3)(1)(-4)}{(7)(2)(5)} = -\frac{12}{70} = -\frac{6}{35}$$

Division:

Division requires that the denominator be inverted and then multiplied by the numerator.

EXAMPLE

$$\frac{\dfrac{2}{3}}{\dfrac{4}{5}} = \frac{2}{3} \div \frac{4}{5} = \frac{2}{3} \times \frac{5}{4} = \frac{10}{12} = \frac{5}{6}$$

Simplest Form:

A fraction can be reduced to simplest form by finding the largest common divisor of both the numerator and denominator. As an illustration,

$$\frac{12}{60} = \frac{12/12}{60/12} = \frac{1}{5}$$

If the largest divisor is not obvious, start with a smaller common divisor, as shown here:

$$\frac{12/6}{60/6} = \frac{2}{10} = \frac{2/2}{10/2} = \frac{1}{5}$$

Or, if you prefer, use repeated divisions of smaller common divisors, such as 2 or 3:

$$\frac{12/2}{60/2} = \frac{6}{30} = \frac{6/2}{30/2} = \frac{3}{15} = \frac{3/3}{15/3} = \frac{1}{5}$$

EXAMPLES

$$\frac{180}{300} = \frac{180/10}{300/10} = \frac{18}{30} = \frac{18/6}{30/6} = \frac{3}{5}$$

$$\frac{72}{256} = \frac{72/8}{256/8} = \frac{9}{32}$$

or

$$\frac{72}{256} = \frac{72/2}{256/2} = \frac{36}{128} = \frac{36/2}{128/2} = \frac{18}{64} = \frac{18/2}{64/2} = \frac{9}{32}$$

Mixed Numbers:

Numbers in mixed form (with whole and fractional parts) can be converted to fractional form by multiplying the whole number by the denominator of the fractional part and adding the numerator.

EXAMPLES

$$3\frac{4}{5} = \frac{(3 \times 5) + 4}{5} = \frac{19}{5}$$

$$60\frac{1}{3} = \frac{(60 \times 3) + 1}{3} = \frac{181}{3}$$

The reverse of this process is simply a division operation with the remainder left in fractional form.

Conversion to Decimal Form:

Fractions can be converted to decimal form by simply performing the indicated division.

EXAMPLES

$$\frac{1}{4} = 4\overline{)1.00}^{\,0.25} = 0.25$$

$$\frac{3}{7} = 7\overline{)3.0000}^{\,0.4285\ldots} = 0.4285\ldots$$

$$4\frac{1}{5} = 4 + 5\overline{)1.0}^{\,0.2} = 4.2$$

EXERCISES

Perform each of the following operations without using a calculator. Be sure all solutions are in simplest form.

1. $\dfrac{3}{5} + \dfrac{1}{3} =$ _____

2. $\dfrac{5}{9} - \dfrac{2}{4} =$ _____

3. $\dfrac{2}{5} - \dfrac{1}{2} + \dfrac{1}{4} =$ _____

4. $\left(\dfrac{5}{6}\right)\left(\dfrac{9}{13}\right) =$ _____

5. $\left(-\dfrac{1}{7}\right)\left(+\dfrac{2}{9}\right)\left(-\dfrac{3}{5}\right) =$ _____

6. $\dfrac{5/6}{2/3} =$ _____

7. $\dfrac{9/10}{-6/7} =$ _____

8. Reduce to simplest form:

    (a) $\dfrac{72}{84} =$ _____

    (b) $\dfrac{144}{384} =$ _____

9. Convert to mixed form:

(a) $\dfrac{16}{7} = $ _____

(b) $\dfrac{320}{9} = $ _____

10. Convert to decimal form using a calculator:

(a) $\dfrac{3}{8} = $ _____

(b) $6\dfrac{5}{6} = $ _____

Powers of 10

The need to work with very small and large numbers requires that the use of powers of 10 be appreciated and clearly understood.

The direction of the shift of the decimal point and the number of places moved will determine the resulting power of 10. For instance:

$$10,000 = 1 \times 10^{+4} = \mathbf{10^{+4}}$$

$$5000 = \mathbf{5 \times 10^{+3}}$$

$$0.000001 = 1 \times 10^{-6} = \mathbf{10^{-6}}$$

$$0.00456 = \mathbf{4.56 \times 10^{-3}}$$

It is particularly important to remember that

$$\boxed{10^0 = 1}$$

(1.1)

EXAMPLES

$$50 \times 10^0 = 50(1) = \mathbf{50}$$

$$0.04 \times 10^0 = 0.04(1) = \mathbf{0.04}$$

Addition and Subtraction:

Addition or subtraction of numbers using scientific notation requires that the power of 10 of each term **be the same.**

EXAMPLES

$$45,000 + 3000 + 500 = 45 \times 10^3 + 3 \times 10^3 + 0.5 \times 10^3 = \mathbf{48.5 \times 10^3}$$

$$0.02 - 0.003 + 0.0004 = 200 \times 10^{-4} - 30 \times 10^{-4} + 4 \times 10^{-4}$$

$$= (200 - 30 + 4) \times 10^{-4} = \mathbf{174 \times 10^{-4}}$$

Multiplication:

Multiplication using powers of 10 employs the following equation, where a and b can be any positive or negative number:

$$10^a \times 10^b = 10^{a+b} \tag{1.2}$$

EXAMPLES

$$(100)(5000) = (10^2)(5 \times 10^3) = 5 \times 10^2 \times 10^3 = 5 \times 10^{2+3} = \mathbf{5 \times 10^5}$$

$$(200)(0.0004) = (2 \times 10^2)(4 \times 10^{-4}) = (2)(4)(10^2)(10^{-4}) = 8 \times 10^{2-4} = \mathbf{8 \times 10^{-2}}$$

Note in the preceding examples that the operations with powers of 10 can be separated from those with integer values.

Division:

Division employs the following equation, where a and b again can be any positive or negative number:

$$\frac{10^a}{10^b} = 10^{a-b} \tag{1.3}$$

EXAMPLES

$$\frac{320{,}000}{4000} = \frac{32 \times 10^4}{4 \times 10^3} = \frac{32}{4} \times 10^{4-3} = 8 \times 10^1 = \mathbf{80}$$

$$\frac{1600}{0.0008} = \frac{16 \times 10^2}{8 \times 10^{-4}} = 2 \times 10^{2-(-4)} = 2 \times 10^{2+4} = \mathbf{2 \times 10^6}$$

Note in the last example the importance of carrying through the proper sign Eq. (1.3).

Powers:

Powers of powers of 10 can be determined using the following equation, where a and b can be any positive or negative number:

$$(10^a)^b = 10^{ab} \tag{1.4}$$

EXAMPLES

$$(1000)^4 = (10^3)^4 = \mathbf{10^{12}}$$

$$(0.002)^3 = (2 \times 10^{-3})^3 = (2)^3 \times (10^{-3})^3 = \mathbf{8 \times 10^{-9}}$$

Scientific Notation

Scientific notation makes use of powers of 10 with restrictions on the mantissa (multiplier) or scale factor (power of the power of 10). It requires that the decimal point appear directly after the first digit greater than or equal to 1 but less than 10.

EXAMPLES

$$2400 = \mathbf{2.40 \times 10^3}$$
$$0.0006 = \mathbf{6.00 \times 10^{-4}}$$
$$5,100,000 = \mathbf{5.1 \times 10^6}$$

Engineering Notation

Engineering notation also makes use of powers of 10 with the restrictions that the power of 10 must be 0 or divisible by 3 and the mantissa must be greater than or equal to 1 but less than 1000.

EXAMPLES

$$22,000 = \mathbf{22.00 \times 10^3}$$
$$0.003 = \mathbf{3.00 \times 10^{-3}}$$
$$0.000065 = \mathbf{65.00 \times 10^{-6}}$$

Conversions Between Levels of Powers of 10

It is often necessary to convert from one power of 10 to another. The process is rather straightforward if you simply keep in mind that an increase or decrease in the power of 10 must be associated with the opposite effect (same whole-number change) on the multiplying factor.

EXAMPLES

Increase by 3

$$160 \times 10^3 = \underline{\qquad} \times 10^6$$

Decrease by 3

$$160 \times 10^3 = \mathbf{0.16 \times 10^6}$$

Decrease by 2

$$0.4 \times 10^{-3} = \underline{\qquad} \times 10^{-5}$$

Increase by 2

$$0.4 \times 10^{-3} = \mathbf{40 \times 10^{-5}}$$

Increase by 4

$$5200 \times 10^{-6} = \underline{\qquad} \times 10^{-2}$$

Decrease by 4

$$5200 \times 10^{-6} = \mathbf{0.52 \times 10^{-2}}$$

EXERCISES

Perform the following operations by hand. Write your answer with the power of 10 indicated.

11. $5,800,000 + 450,000 + 2000 =$ _____ $\times 10^3$

12. $0.04 + 0.008 - 0.3 =$ _____ $\times 10^{-3}$

13. $2400 + 0.05 \times 10^3 - 40,000 \times 10^{-3} =$ _____ $\times 10^3$

14. $(68,000)(40,000) =$ _____ $\times 10^9$

15. $(0.0009)(0.006) =$ _____ $\times 10^{-6}$

16. $(-5 \times 10^{-8})(8 \times 10^5)(-0.02 \times 10^4) =$ _____ $\times 10^0$

17. $\dfrac{0.00081}{0.009} =$ _____ $\times 10^{-2}$

18. $\dfrac{5000}{6 \times 10^6} =$ _____ $\times 10^{-4}$

19. $\dfrac{-8 \times 10^{-4}}{0.002} =$ _____ $\times 10^{-1}$

20. $(4 \times 10^5)^3 =$ _____ $\times 10^{16}$

21. $(0.0003)^4 =$ _____ $\times 10^{-15}$

22. $[(4000)(0.0003)]^3 =$ _____ $\times 10^0$

Prefixes

The frequent use of some powers of 10 has resulted in their being assigned abbreviations that can be applied as prefixes to a numerical value in order to quickly identify its relative magnitude.

The following is a list of the most frequently used prefixes in electrical and electronic engineering technology:

$$10^{12} = \textbf{tera (T)}$$
$$10^9 = \textbf{giga (G)}$$
$$10^6 = \textbf{mega (M)}$$
$$10^3 = \textbf{kilo (k)}$$
$$10^{-3} = \textbf{milli (m)}$$
$$10^{-6} = \textbf{micro } (\boldsymbol{\mu})$$
$$10^{-9} = \textbf{nano (n)}$$
$$10^{-12} = \textbf{pico (p)}$$

EXAMPLES

$$6,000,000 \, \Omega = 6 \times 10^6 \, \Omega = \textbf{6 M}\boldsymbol{\Omega}$$
$$0.04 \, A = 40 \times 10^{-3} \, A = \textbf{40 mA}$$
$$0.00005 \, V = 50 \times 10^{-6} \, V = \textbf{50 } \boldsymbol{\mu}\textbf{V}$$

When converting from one form to another, be aware of the relative magnitude of the quantity before you start to provide a check once the conversion is complete. Starting out with a relatively small quantity and ending up with a result of large relative magnitude clearly indicates an error in the conversion process.

The most direct method of converting from one power of 10 to another is to write the number in the power-of-10 format and note the requested change in the power of 10. If the power of 10 is larger, the decimal point of the multiplier must be moved to the left a number of places equal to the increase in the power of 10. If the requested power of 10 is smaller, the decimal point is moved to the right a number of places equal to the decrease in the power of 10.

EXAMPLES

(a) Convert 0.008 MΩ to kilohms.

Decrease by 3

$$0.008 \times 10^{+6} \, \Omega \longrightarrow 0.008. \times 10^{+3} \, \Omega = \textbf{8 k}\boldsymbol{\Omega}$$

Increase by 3

(b) Convert 5600 mA to amperes.

Increase by 3

$$5600 \times 10^{-3} \, A \longrightarrow 5.600. \times 10^{0} \, A = \textbf{5.6 A}$$

Decrease by 3

EXERCISES

Apply the most appropriate prefix to each of the following quantities.

23. 0.00006 A = _____

24. 1,504,000 Ω = _____

25. 32,000 V = _____

26. 0.000000009 A = _____

Perform the following conversions.

27. 35 mA = _____A

28. 0.005 kV = _____ V = _____ mV

29. 8,500,000 Ω = _____ MΩ = _____ kΩ

30. 4000 pF = _____ nF = _____ μF

Square and Cube Roots

When determining the square or cube root of a number by hand, the power of 10 must be divisible by 2 or 3, respectively, or the resulting power of 10 will not be a whole number.

EXAMPLES

$$\sqrt{200} = (200)^{1/2} = (2 \times 10^2)^{1/2} = (2)^{1/2} \times (10^2)^{1/2} \cong 1.414 \times 10^1 = \mathbf{14.14}$$

$$\sqrt{0.004} = (0.004)^{1/2} = (40 \times 10^{-4})^{1/2} = (40)^{1/2} \times (10^{-4})^{1/2}$$

$$= 6.325 \times 10^{-2} = \mathbf{0.06325}$$

$$\sqrt[3]{5000} = (5000)^{1/3} = (5 \times 10^3)^{1/3} = (5)^{1/3} \times 10^1 \cong 1.71 \times 10^1 = \mathbf{17.1}$$

Some calculators provide the $\sqrt[x]{}$ function, whereas others require that you use the $\sqrt[x]{y}$ function. For some calculators, the y^x function is all that is available, requiring that the y value be entered first, followed by the y^x function and then the power ($x = 1/3 = 0.3333$).

The use of a calculator removes the need to worry about the power of 10 under the radical sign.

EXERCISES

Use a calculator.

31. $\sqrt{6.4}$ = _____

32. $\sqrt[3]{3000}$ = _____

33. $\sqrt{0.00007}$ = _____

Exponential Functions

The exponential functions e^x and e^{-x} will appear frequently in the analysis of R-C and R-L networks with switched dc inputs (or square-wave inputs). On most calculators, only e^x is provided, requiring that the user insert the negative sign when necessary.

Keep in mind that e^x is equivalent to $(2.71828 \ldots)^x$ or a number to a power x. For any positive values of x greater than 1, the result is a magnitude greater than $2.71828 \ldots$. In other words, for increasing values of x, the magnitude of e^x will increase rapidly. For $x = 0$, $e^0 = 1$, and for increasing negative values of x, e^{-x} will become increasingly smaller.

When inserting the negative sign for e^{-x} functions, be sure to use the proper key to enter the negative sign.

EXAMPLES

$$e^{+2} \cong 7.3890$$

$$e^{+10} \cong 22{,}026.47$$

$$e^{-2} \cong 0.13534$$

$$e^{-10} \cong 0.0000454$$

Also keep in mind that

$$e^{-x} = \frac{1}{e^x} \text{ and } e^{+x} = \frac{1}{e^{-x}}$$

(1.5)

EXERCISES

Use a calculator.

34. $e^{+4.2} =$ _____

35. $e^{+0.5} =$ _____

36. $e^{-0.02} =$ _____

37. $\dfrac{1}{e^{-3}} =$ _____

Algebraic Manipulations

The following is a brief review of some basic algebraic manipulations that must be performed in the analysis of dc circuits. It is assumed that a supporting math course will expand on the coverage provided here.

EXAMPLES

(a) Given $v = \dfrac{d}{t}$, solve for d and t.

Both sides of the equation must first be multiplied by t:

$$(t)(v) = \left(\frac{d}{t}\right)(t)$$

resulting in $d = vt$.

Dividing both sides of $d = vt$ by yields

$$\frac{d}{v} = \frac{vt}{v}$$

$$t = \frac{d}{v}$$

or

In other words, the proper choice of multiplying factors for both sides of the equation will result in an equation for the desired quantity.

(b) Given $R_1 + 4 = 3R_1$, solve for R_1.

In this case, $-R_1$ is added to both sides, resulting in

$$(R_1 + 4) - R_1 = 3R_1 - R_1$$

or
$$4 = 2R_1$$

so
$$R_1 = \frac{4}{2} = 2$$

(c) Given $\dfrac{1}{R_1} = \dfrac{1}{3R_1} + \dfrac{1}{6}$, solve for R_1.

Multiplying both sides of the equation by R_1 results in

$$\frac{\not{R_1}}{\not{R_1}} = \frac{\not{R_1}}{3\not{R_1}} + \frac{R_1}{6}$$

or
$$1 = \frac{1}{3} + \frac{R_1}{6}$$

Subtracting 1/3 from both sides gives

$$1 - \frac{1}{3} = \frac{1}{3} - \frac{1}{3} + \frac{R_1}{6}$$

or
$$\frac{2}{3} = \frac{R_1}{6}$$

so
$$R_1 = \frac{(6)(2)}{3} = \frac{12}{3} = 4 \, \Omega$$

EXERCISES

Show all work in an organized and neat manner.

38. Given $P = I^2R$, solve for R and I.

$R = $ _____, $I = $ _____

39. Given $30I = 5I + 5$, solve for I.

$I =$ _____

40. Given $\dfrac{1}{R_T} = \dfrac{1}{R_1} + \dfrac{1}{4R_1}$, solve for R_T in terms of R_1.

$R_T =$ _____

41. Given $F = \dfrac{kQ_1Q_2}{r^2}$, solve for Q_1 and r.

$Q_1 =$ _____, $r =$ _____

CALCULATOR FUNDAMENTALS

Calculators have become such a fundamental and important part of the technology program that every effort must be made to ensure the basic operations are correctly understood. Too often, the manual associated with a calculator is never read and the student assumes the operations he or she needs can be implemented easily simply by examining the keypad. The result, however, is that important facets of the calculator, such as the hierarchy of operations, the proper input of data, and so forth are learned only after disaster has occurred once or twice.

The purpose of this introduction, therefore, is to discuss only the very important elements of calculator use to ensure the operations required in the dc section of this laboratory manual can be performed correctly. Most of the content of this section applies to any handheld calculator.

Initial Settings

Every scientific calculator has a mode setting whereby the two important choices of format and level of accuracy can be made. Consult your manual for the manner in which the format and level of accuracy are chosen.

Format:

The chosen format determines how the numbers will appear after any of the fundamental operations are performed, such as addition, subtraction, multiplication, or square root. In other words, the numbers can be entered in *any* format, but the result will appear in the *chosen* format. For most scientific calculators, the choices are **normal, scientific,** and **engineering.** Examples of each are shown for the basic operation of dividing 1 by 3.

$$\textbf{Normal:} \quad 1/3 = 0.33$$
$$\textbf{Scientific:} \quad 1/3 = 3.33E\text{-}1$$
$$\textbf{Engineering:} \quad 1/3 = 333.33E\text{-}3$$

Note that the normal format simply places the decimal point in the most logical location. The scientific ensures that the number preceding the decimal point is a single digit followed by the required power of 10. The engineering format will always ensure that the power of 10 is a multiple of 3 (whether it is positive, negative, or zero). Because specific names have been given to powers of 10 that are multiples or divisible by 3 in the engineering format, it will be the choice for the greater part of this manual. Once you have entered a number in the normal format, you can easily obtain the scientific or engineering format first by choosing the SCI or ENG format and then pressing the Enter key for some calculators or simply dividing or multiplying by 1 for others.

$$0.003 = \textbf{3.00E-3}$$
$$45{,}000 = \textbf{45.00E3}$$
$$0.000000000003 = \textbf{3.00E-12}$$

Accuracy:

In the preceding examples, the numbers are accurate to the hundredths place. Your instructor may prefer a different level of accuracy. Whatever the choice, take the time now to learn how to set the level of accuracy so you don't end up with lengthy numbers whose level of accuracy makes no sense whatsoever for the operations to be performed.

EXERCISES

Using a calculator, convert the following from the normal to scientific format using hundredths place accuracy.

42. $0.000045 =$ _____

43. $3400000 =$ _____

44. $0.08 =$ _____

45. $220 =$ _____

46. $0.00000001 =$ _____

Convert to engineering format using hundredths place accuracy.

47. $0.000045 =$ _____

48. $3400000 =$ _____

49. $0.08 =$ _____

50. $220 =$ _____

51. $0.00000001 =$ _____

Order of Operations

If there is one element of using a calculator that is taken too lightly, it is the order in which to perform the operations appearing in an equation.

For instance, the equation

$$\frac{8}{3 + 1}$$

is often entered as

$$\boxed{8} \quad \boxed{\div} \quad \boxed{3} \quad \boxed{+} \quad \boxed{1} \quad \Rightarrow \frac{8}{3} + 1 = 2.67 + 1 = 3.67$$

which is **totally incorrect** (2 is the correct answer).

You need to be aware that a calculator *will not* perform the addition first and then the division. In fact, addition and subtraction are the last operations to be performed in any equation. It is therefore very important that you carefully read and understand the next few paragraphs if you are to use your calculator properly.

1. The first operations to be performed by the calculator can be set using **parentheses ( )**. It does not matter which operations are within the parentheses. They dictate to the calculator that these operations are to be performed first. There is no limit on the number of parentheses used in each equation—they will all show operations to be performed first. For instance, for the preceding example, if parentheses are added as shown next, the addition will be performed first and the correct answer will be obtained.

$$\frac{8}{(3 + 1)} \Rightarrow \boxed{8} \quad \boxed{\div} \quad \boxed{(} \quad \boxed{3} \quad \boxed{+} \quad \boxed{1} \quad \boxed{)} \Rightarrow \frac{8}{4} = \mathbf{2}$$

2. Next, **powers and roots** (other than those appearing as a defined key, such as $\sqrt{}$) are performed such as x^3, $\sqrt[3]{2}$, and so forth.

3. **Negation** (applying a negative sign to a quantity) and **single-key operations,** such as $\sqrt{}$, sin, and $\tan^{-1}$ are performed.

4. **Multiplication** and **division** are then performed.

5. Lastly, **addition** and **subtraction** are performed.

A compressed visual review is provided below:

( )   *First*

x^2

$-$ **(to a quantity)**

sin

x, $\div$

$+$, $-$ *Last*

It may take a few moments and some repetition to remember this order, but at least you are aware that there is some order to the operations and that ignoring them can result in some disastrous conclusions.

EXAMPLES

(a) Determine

$$\sqrt{\frac{9}{3}}$$

The following calculator operations will result in an **incorrect** answer of 1 because the square root operation will be performed before the division.

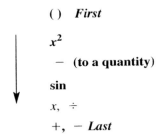 $\Rightarrow \dfrac{\sqrt{9}}{3} = \dfrac{3}{3} = 1.00$

However, recognizing that we must first divide 9 by 3, we can use parentheses as follows so this operation is the first to be performed and the correct answer is obtained.

$\boxed{\sqrt{}}\ \boxed{(}\ \boxed{9}\ \boxed{\div}\ \boxed{3}\ \boxed{)} \Rightarrow \sqrt{\left(\dfrac{9}{3}\right)} = \sqrt{3} = \mathbf{1.67}$

(b) Determine

$$\frac{8 + 12}{2}$$

If you simply entered what is shown next, you will obtain an **incorrect** result, because the division is performed first.

$\boxed{8}\ \boxed{+}\ \boxed{1}\ \boxed{2}\ \boxed{\div}\ \boxed{2} \Rightarrow 8 + \dfrac{12}{2} = 8 + 6 = 14.00$

Using parentheses results in the correct answer.

$\boxed{(}\ \boxed{8}\ \boxed{+}\ \boxed{1}\ \boxed{2}\ \boxed{)}\ \boxed{\div}\ \boxed{2} \Rightarrow \dfrac{(8 + 12)}{2} = \dfrac{20}{2} = \mathbf{10}$

(c) Find

$$\sqrt{2^2 + 4^2}$$

In this case, you should square the numbers first, add, and then take the square root. If you enter the following, you will obtain an **incorrect** answer.

$$\boxed{\sqrt{}}\ \boxed{2}\ \boxed{x^2}\ \boxed{+}\ \boxed{4}\ \boxed{x^2} \Rightarrow \sqrt{2^2} + 4^2 = 2 + 16 = 18.00$$

However, if you enter the following sequence, you will get the correct answer.

$$\boxed{\sqrt{}}\ \boxed{(}\ \boxed{2}\ \boxed{x^2}\ \boxed{+}\ \boxed{4}\ \boxed{x^2}\ \boxed{)} \Rightarrow \sqrt{(2^2 + 4^2)} = \sqrt{(4 + 16)} = \sqrt{20} = \mathbf{4.47}$$

(d) Calculate

$$\frac{6}{2 + 5}$$

If you enter this sequence, an incorrect answer of 8 will result

$$\boxed{6}\ \boxed{\div}\ \boxed{2}\ \boxed{+}\ \boxed{5} \Rightarrow \frac{6}{2} + 5 = 3 + 5 = 8$$

Using brackets to ensure the addition takes place before the division will result in the correct answer:

$$\boxed{6}\ \boxed{\div}\ \boxed{(}\ \boxed{2}\ \boxed{+}\ \boxed{5}\ \boxed{)} \Rightarrow \frac{6}{(2 + 5)} = \frac{6}{7} = \mathbf{0.86}$$

(e) Determine

$$\frac{2 + 3}{6^2}$$

The first operation to be performed is the squaring of the number six in the denominator, which is fine for the answer desired. However, since division is the next operation to be performed, 6^2 will be divided into 3, resulting in an **incorrect** answer of 2.08.

$$\boxed{2}\ \boxed{+}\ \boxed{3}\ \boxed{\div}\ \boxed{6}\ \boxed{\wedge}\ \boxed{2} \Rightarrow 2 + \frac{3}{6^2} = 2 + 0.08 = 2.08$$

Using parentheses will ensure that the addition takes place first and the correct answer is obtained:

$$\boxed{(}\ \boxed{2}\ \boxed{+}\ \boxed{3}\ \boxed{)}\ \boxed{\div}\ \boxed{6}\ \boxed{\wedge}\ \boxed{2} \Rightarrow \frac{(2 + 3)}{6^2} = \frac{5}{36} = \mathbf{0.14}$$

(f) Determine the sum

$$\frac{1}{4} + \frac{1}{6} + \frac{2}{3}$$

Because the division will occur first, the correct result will be obtained by simply performing the operations as indicated:

$$\boxed{1}\ \boxed{\div}\ \boxed{4}\ \boxed{+}\ \boxed{1}\ \boxed{\div}\ \boxed{6}\ \boxed{+}\ \boxed{2}\ \boxed{\div}\ \boxed{3} \Rightarrow \mathbf{1.08}$$

(g) Find the cube root of 3000:

$$\sqrt[3]{3000} = (3000)^{1/3}$$

Always keep in mind that the square root is the same as taking a quantity to the 1/2 power, the cube root is to the 1/3 power, the fourth root is to the 1/4 power, and so on. Remember, however, that the power is taken before the division, so a parentheses must surround the division:

$$\boxed{3}\ \boxed{0}\ \boxed{0}\ \boxed{0}\ \boxed{\wedge}\ \boxed{(}\ \boxed{1}\ \boxed{\div}\ \boxed{3}\ \boxed{)} \Rightarrow \mathbf{14.42}$$

EXERCISES

Determine the following using a calculator.

52. $\dfrac{4 + 6}{2 + 8} =$ _____

53. $\dfrac{2 \times (8 + 6)}{10} =$ _____

54. $\dfrac{4 + 9}{6 \times 2} =$ _____

55. $\dfrac{4^3}{2(1 + 2)} =$ _____

56. $\dfrac{1}{6 + 3(2 + 3)} =$ _____

57. $\sqrt{(2 + 3)^2 + 4^2} =$ _____

58. $\dfrac{1}{(3 + 1)^2} =$ _____

59. $\dfrac{\sqrt[3]{5^2}}{3^2 - 2} =$ _____

Powers of 10

Powers of 10 are entered using the E, EE, or EXP key on your calculator. Always remember, however, to use the negative sign on the numeric keypad when entering a negative power. *Do not use the subtraction key.* For all operations, negative numbers are entered with the negative sign on the numeric keypad. The negative sign for subtraction is used only for that specific operation. The positive sign associated with positive powers is understood and does not appear on the display.

With the calculator set for the SCI and ENG mode, the proper power of 10 for any number can be found by simply entering the number in the decimal form and pressing the Enter key for some calculators or dividing or multiplying by 1 for others. For the ENG mode with hundredths place accuracy, the following will result:

$$0.006 = \mathbf{6.00E-3}$$

$$80400 = \mathbf{80.40E3}$$

This sequence is also applicable to numbers in the power-of-10 format. Each number in the power-of-10 format is treated as a single digit in a defined sequence.

EXAMPLES

$$0.04 + \frac{8}{3000} = 40E-3 + \left(\frac{8}{3E3}\right)$$

$$= \boxed{4}\ \boxed{0}\ \boxed{EE}\ \boxed{-}\ \boxed{3}$$

$$\boxed{+}\ \boxed{(}\ \boxed{8}\ \boxed{\div}\ \boxed{3}\ \boxed{EE}\ \boxed{3}\ \boxed{)} \Rightarrow \mathbf{42.67E-3}$$

$$(0.003)^2 \times 4000 = (3E-3)^2 \times 4E3$$

$$= \boxed{3}\ \boxed{EE}\ \boxed{-}\ \boxed{3}\ \boxed{x^2}\ \boxed{\times}\ \boxed{4}\ \boxed{EE}\ \boxed{3} \Rightarrow \mathbf{36.00E-3}$$

EXERCISES

Use a calculator.

60. $50E3 - 2.2E4 + 0.08E6 = $ _____

61. $\dfrac{(4 \times 10^{-6})^2}{8 \times 10^3} = $ _____

62. $0.008 \times (6 \times 10^3) + 5200 \times 10^{-3} - 200E-3 - 3 = $ _____

63. $80{,}000 \times \sqrt{(2000)^3} = $ _____

Exponential Functions

The exponential function appears on all scientific calculators because it is such a prominent element of many important equations. When e^x is chosen, the display will appear with a request for the power (x). As before, be sure to follow the correct sequence of operations if the power is the result of a mathematical operation. In addition, be sure to use the negative sign on the numeric keyboard for negative powers.

EXAMPLES

(a) Determine $e^{6/2} = e^{(6/2)}$.

| 2nd | e^x | (| 6 | ÷ | 2 |) | = 20.09E0 = **20.09**

For the TI-89 calculator:

| ◇ | e^x | 6 | ÷ | 2 |) | ENTER | = **20.1E0**

(b) Find $\dfrac{3}{e^{1+4/3}} + e^{-2} = \dfrac{3}{e^{(1+4/3)}} + e^{-2}$.

| (| 3 | ÷ | 2nd | e^x | (| 1 | + | 4 | ÷ | 3 |) |)

| + | 2nd | e^x | − | 2 | = **426.25E−3**

For the TI-89 calculator:

| (| 3 | ÷ | ◇ | e^x | (| 1 | + | 4 | ÷ | 3 |)

|) | + | ◇ | (−) | 2 |) |) | ENTER | = **426.E−3**

EXERCISES

Use a calculator.

64. $e^{\sqrt{10/3}} = $ _____

65. $e^{e^2} = $ _____

66. $\dfrac{68E3 \times e^{-1}}{4E3 + e^8} = $ _____

The primary purpose of this section of the laboratory manual is to ensure your understanding of the importance of being aware of which calculations are performed first when using a calculator. Too often there is no awareness of the order of operations and ridiculous results are accepted as correct because the entry is carefully checked and everything looks fine. The same priority of operations discussed here applies to most computers, so the carryover is almost universal.

Resistors and the Color Code

OBJECTIVES

1. Become familiar with the digital and analog ohmmeter.
2. Learn to read and use the resistor color code.
3. Become aware of the magnitude and impact of the internal resistance of a voltmeter and ammeter.

EQUIPMENT REQUIRED

Resistors

1—1-MΩ, 1-W film resistor

1—1-MΩ, 2-W film resistor

1—1-MΩ, 1/2-W film resistor

1—6.8-Ω, 91-Ω, 220-Ω, 3.3-kΩ, 10-kΩ, 470-kΩ, 1-MΩ, 1/4-W film resistors

Instruments

1—VOM (Volt-Ohm-Milliammeter)

1—DMM (Digital Multimeter)

EQUIPMENT ISSUED

TABLE 2.0

Item	Manufacturer and Model No.	Laboratory Serial No.
VOM		
DMM		

DESCRIPTION OF EQUIPMENT

Both the VOM and the DMM will be used in this experiment to measure the resistance of the provided resistors. The VOM employs an analog scale to read resistance, voltage, and current, whereas the DMM has a digital display. An analog scale is a continuous scale, requiring that the user be able to interpret the location of the pointer using the scale divisions provided. For resistance measurements, the analog scale is also nonlinear, resulting in smaller distances between increasing values of resistance. Note on the VOM scale the relatively large distance between 1 Ω and 10 Ω and the smaller distance between 100 Ω and 1000 Ω. The digital display provides a numerical value with the accuracy determined by the chosen scale. For many years the analog meter was the instrument employed throughout the industry. In recent years, the digital meter has grown in popularity, making it important that the graduate of any technical program be adept at using both types of meters.

RÉSUMÉ OF THEORY

In this experiment, the resistance of a series of 1/2-W film resistors first will be determined from the color code and then compared with the measured value using both the VOM and the DMM. The two meters were applied to ensure familiarity with reading both an analog and a digital scale.

The procedure for determining the resistance of a color-coded resistor is described in Appendix I with a listing of the numerical value associated with each color. The numerical value associated with each color has been repeated in Fig. 2.1. The first two bands (those closest to the end of the resistor) determine the first two digits of the resistor value, while the third band determines the power of the power-of-10 multiplier (actually the number of zeros to follow the first two digits). If the third band is silver (0.01) or gold (0.1), it is a multiplying factor used to establish resistor values less than 10 Ω. The fourth band is the percent tolerance for the chosen resistor (see Fig. 2.1).

Black	Brown	Red	Orange	Yellow	Green	Blue	Purple	Gray	White
0	**1**	**2**	**3**	**4**	**5**	**6**	**7**	**8**	**9**

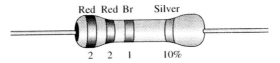

Red Red Br   Silver

2   2   1   10%

FIG. 2.1   220 Ω ± 10% = 220 Ω ± 22 Ω = 198 Ω to 242 Ω

For increasing wattage ratings, the size of the film resistor will increase to provide the "body" required to dissipate the resulting heating effects.

Resistance is *never* measured by an ohmmeter in a *live* network, due to the possibility of damaging the meter with excessively high currents and obtaining readings that have no meaning. It is usually best to remove the resistor from the circuit before measuring its resistance to ensure that the measured value does not include the effect of other resistors in the system. However, if this maneuver is undesirable or impossible, make sure that one end of the resistor is not connected to any other element.

The use of the VOM and DMM will be described early in the laboratory session. For the VOM, always reset the zero-adjust whenever you change scales. In addition, always choose the dial setting (R × 1, R × 10, and so on) that will place the pointer in the region of the scale that will give the best reading. Finally, do not forget to multiply the reading by the proper multiplying factor. For the DMM, remember that any scale marked "kΩ" will be reading in kilohms, and any "MΩ" scale in megohms. For instance, a 91-Ω resistor may read 90.7 Ω on a 200-Ω scale, 0.091 on a 2-kΩ scale, and 0.000 on a 2-MΩ scale. There is no zero-adjust on a DMM meter, but make sure that $R = 0\ \Omega$ when the leads are touching, or else an adjustment internal to the meter may have to be made. Any resistance above the maximum for a chosen scale will result in an O.L. indication.

It is important to remember that there is no polarity to resistance measurements. Either lead of the meter can be placed on either end of the resistor—the resistance will be the same. Polarities will become important, however, when we measure voltages and currents in the experiments to follow.

PROCEDURE

The purpose of this first experiment is to acquaint you with the laboratory equipment, so *do not rush*. Learn how to read the meter scales accurately, and take your data carefully. If you are uncertain about anything, do not hesitate to ask your instructor.

Part 1 Body Size

(a) In the space below, draw the physical sizes of 1/4-W, 1/2-W, 1-W, and 2-W, 1-MΩ film resistors (color bands = brown, black, green). Note in particular that the resistance of each is the same but that the size increases with wattage rating. Identify each by its wattage rating.

(b) How much larger (physically) is the 1-W resistor than the 1/2-W resistor? Is the ratio about the same for the 2-W resistor as compared with the 1-W resistor? Answer the questions in sentence form.

Part 2   Color Code

(a) Using the procedure described in the Résumé of Theory, determine the color bands for each resistor appearing in Table 2.1. Then find each resistor in your provided kit and enter the colors for all four bands in Table 2.1, as shown by the example.

(b) Enter the numerical value of each color in the next column, as shown by the example.

TABLE 2.1

Resistor (Nominal Value)	Color Bands—Color				Color Bands—Numerical Value			
	1	2	3	4	1	2	3	4
22 Ω	Red	Red	Black	Gold	2	2	0	5%
91 Ω								
220 Ω								
3.3 kΩ								
10 kΩ								
470 kΩ								
1 MΩ								
6.8 Ω								

(c) The percent tolerance is used to determine the range of resistance levels within which the manufacturer guarantees the resistor will fall. It is determined by first taking the percent tolerance and multiplying by the nominal resistance level. For the example in Table 2.1, the resulting resistance level

$$(5\%)(22 \ \Omega) = (0.05)(22 \ \Omega) = 1.1 \Omega$$

is added to and subtracted from the nominal value to determine the range as follows:

$$\text{Maximum value} = 22 \ \Omega + 1.1 \ \Omega = 23.1 \ \Omega$$
$$\text{Minimum value} = 22 \ \Omega - 1.1 \ \Omega = 20.9 \ \Omega$$

as shown in Table 2.2.

Complete Table 2.2 for each resistor in Table 2.1.

(d) Read the resistance level using the DMM and insert the value in Table 2.3. For each resistor, take the time to choose the scale that will result in the highest degree of accuracy for the measurement. Ask for assistance if you need it.

Then determine the magnitude of the difference between the nominal and measured values using the following equation. The vertical bars of the equation specify that the sign resulting from the operation is not to be included in the solution.

TABLE 2.2

Resistor	Minimum Resistance	Maximum Resistance
22 Ω	20.9 Ω	23.1 Ω
91 Ω		
220 Ω		
3.3 kΩ		
10 kΩ		
470 kΩ		
1 MΩ		
6.8 Ω		

$$\% \text{ Difference} = \left| \frac{\text{Nominal} - \text{Measured}}{\text{Nominal}} \right| \times 100\% \qquad (2.1)$$

TABLE 2.3

Meter	DMM			VOM		
Normal Resistor Value	Measured Value	Falls within Specified Tolerance (Yes/No)	% Difference	Measured Value	Falls within Specified Tolerance (Yes/No)	% Difference
Sample: 22 Ω	22.9 Ω	Yes	4.09%	23 Ω	Yes	4.5%
91 Ω						
220 Ω						
3.3 kΩ						
10 kΩ						
470 kΩ						
1 MΩ						
6.8 Ω						

(e) Repeat part (d) using the VOM and insert the data in Table 2.3. As with the DMM, be sure to ask for assistance if you need it. It is very important to become familiar with the operation of each meter.

(f) For the DMM measurements, are all the resistors within the specified tolerance range? You can check by referencing Table 2.2 or comparing the percent tolerance to that listed in Table 2.1. If not, how many resistors did not fall within the tolerance range?

(g) For the VOM measurements, are all the resistors within the specified tolerance range? Again, you can check using the range of Table 2.2 or the percent tolerance of Table 2.1. If not, how many resistors did not fall within the tolerance range?

Part 3   Body Resistance

Guess the resistance of your body between your hands and record the value in Table 2.4. Measure the resistance with the DMM by firmly holding one lead in each hand (wet your fingers and hold the leads as tight as possible), and record in the same table. If 10 mA are "lethal," what voltage ($V = IR$) would be required to produce the current through your body? Again, record in Table 2.4.

TABLE 2.4

Guessed body resistance	
Measured body resistance	
Lethal voltage	

Calculation:

Part 4   Meter Resistance

(a) Voltmeters Ideally, the internal resistance of the DMM and VOM should be infinite (like an open circuit) when voltages in a network are being measured to ensure that the meter does not alter the normal behavior of the network.

For the VOM, there is an ohm/volt (Ω/V) rating written on the bottom of the scale that permits determination of the internal resistance of each scale of the meter when used as a voltmeter. For the 260 Simpson VOM, the ohm/volt rating is 20,000 Ω/V. The internal resistance of each setting then can be calculated by multiplying the maximum voltage reading of a scale by the ohm/volt rating. For instance, the 10-V scale will have (10 V)(20,000 Ω/V) = 200 kΩ, whereas the 250-V scale will have $(250\ \text{V})(20{,}000\ \Omega/\text{V}) = 5\ \text{M}\Omega$. **No matter what voltage level is measured on each scale, the internal resistance will remain the same.**

Using the DMM as an ohmmeter, measure the resistance of each dc voltage scale of the VOM and record in Table 2.5. Complete the table only for those scales found on your VOM. If the calculated resistance is greater than the range of your DMM, simply put a dash in the space provided for the measured resistance.

TABLE 2.5

VOM	Ω/V Rating	Calculated Resistance	Measured Resistance	% Difference = $\left\|\dfrac{\text{Calc.} - \text{Meas.}}{\text{Calc.}}\right\| \times 100\%$
2.5 V				
10 V				
50 V				
250 V				
1000 V				

Do the resulting percent differences verify that the internal resistance of a VOM can be determined using the procedure just outlined? Answer the question in sentence form.

Most DMMs have the same internal resistance for all the dc voltage scales. There are a few meters, however, that have a lower internal resistance for some lower scales, such as less than 1 V. An internal resistance of 10 MΩ to 20 MΩ is typical for a variety of commercially available DMMs.

Use your neighbor's DMM to measure the internal resistance of each dc voltage scale of your DMM and complete Table 2.6 for all the scales that appear on your meter. If the internal

TABLE 2.6

DMM	Specified Internal Resistance	Measured Resistance
200 mV		
2 V		
20 V		
200 V		
1500 V		

resistance is more than the maximum resistance the DMM can read, simply insert the maximum possible reading for your DMM in the measured resistance column.

Based on the fact that the internal resistance of a meter should be as large as possible for each dc voltage scale, which meter (the VOM or DMM) would appear to disturb a network the least for most voltage scales? Answer in sentence form with some specifics.

(b) Ammeters The ammeter is also an instrument that when inserted in a network should not adversely affect the normal current levels. However, since it is placed in series with the branch in which the current is being measured, its resistance should be as small as possible.

Using the DMM as an ohmmeter, measure the resistance of each dc current scale of the VOM and record in Table 2.7. Change the maximum value for each current scale of Table 2.7 if different on your VOM.

TABLE 2.7

VOM	Measured Resistance
1 mA	
10 mA	
100 mA	
500 mA	

Using the VOM as an ohmmeter, measure the resistance of each current scale of the DMM and record in Table 2.8. Change the maximum value for each current scale of Table 2.8 if different for your DMM.

TABLE 2.8

DMM	Measured Resistance
2 mA	
20 mA	
200 mA	
2 A	

Based on the fact that the internal resistance of an ammeter should be as small as possible for all current ranges, which meter (the VOM or DMM) would appear to disturb the network the least? Answer the question in sentence form.

EXERCISES

1. What are the ohmic values and tolerances of the following commercially available carbon resistors?

Color Bands—Color				Numerical Value	Tolerance
1	**2**	**3**	**4**		
Brown	Black	Blue	Gold		
Yellow	Violet	Orange	Gold		
Brown	Gray	Gold	None		
Red	Yellow	Silver	Gold		
Green	Brown	Green	Silver		
Green	Blue	Black	None		

2. For the VOM, which region of the scale normally provides the best readings? Why?

3. In your own words, review the procedure for using a DMM to read the resistance of a resistor.

4. In your own words, review the procedure for using a VOM to read the resistance of a resistor.

5. Based on your laboratory experience, what are the relative advantages of a digital DMM meter over an analog VOM meter?

6. Under what conditions would an analog display have advantages over a digital display?

Ohm's Law

OBJECTIVES

1. Become familiar with the dc power supply and setting the output voltage.
2. Measure the current in a dc circuit.
3. Apply and plot Ohm's law.
4. Determine the slope of an *I-V* curve.
5. Become more familiar with the use of the analog VOM and digital DMM.

EQUIPMENT REQUIRED

Resistors

1—1-kΩ, 3.3-kΩ (1/4-W)

Instruments

1—DMM
1—VOM
1—dc power supply

EQUIPMENT ISSUED

TABLE 3.0

Item	Manufacturer and Model No.	Laboratory Serial No.
DMM		
VOM		
Power supply		

RÉSUMÉ OF THEORY

In any active circuit there must be a source of power. In the laboratory, it is convenient to use a source that requires a minimum of maintenance and, more important, whose output voltage can be varied easily. Power supplies are rated as to maximum voltage and current output. For example, a supply rated 0–40 V at 500 mA will provide a maximum voltage of 40 V and a maximum current of 500 mA at any voltage.

Most dc power supplies have three terminals, labeled as shown in Fig. 3.1. The three terminals permit the establishment of a positive or negative voltage, which can be grounded or ungrounded. The variable voltage is available only between terminals A and B. Both A and B must be part of any connection scheme. If only terminals A and B are employed, as shown in Fig. 3.2, the supply is considered "floating" and not connected to the common ground of the network. For common ground and safety reasons, the supply is normally grounded as shown in Fig. 3.3 for a positive voltage and as in Fig. 3.4 for a negative voltage.

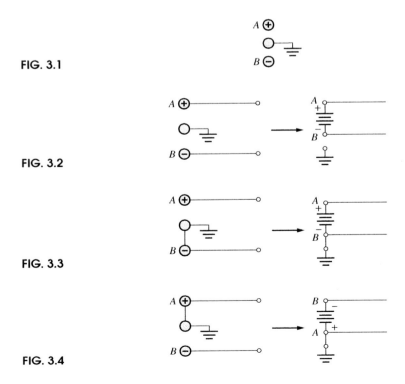

FIG. 3.1

FIG. 3.2

FIG. 3.3

FIG. 3.4

When measuring voltage levels, make sure the voltmeter is connected in parallel (across) the element being measured, as shown in Fig. 3.5. In addition, recognize that if the leads are connected as shown in the figure, the reading will be up-scale and positive. If the meter were hooked up in the reverse manner, a negative (down-scale, below-zero) reading would result. The voltmeter is therefore an excellent instrument not only for measuring the voltage level but also for determining the polarity. Since the meter is always placed in parallel with the element, there is no need to disturb the network when the measurement is made.

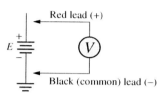

FIG. 3.5

Ammeters are always connected in series with the branch in which the current is being measured, as shown in Fig. 3.6, normally requiring that the branch be opened and the meter inserted. Ammeters also have polarity markings to indicate the manner in which they should be connected to obtain an up-scale reading. Since the current I of Fig. 3.6 would establish a voltage drop across the ammeter as illustrated, the reading of the ammeter will be up-scale and positive. If the meter were hooked up in the reverse manner, the reading would be negative or down-scale. In other words, simply reversing the leads will change a below-zero indication to an up-scale reading.

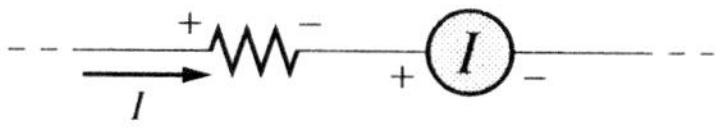

FIG. 3.6

Until you become familiar with the use of the ammeter, draw in the ammeter in the network with the polarities determined by the current direction. It is then easier to ensure that the meter is connected properly to the surrounding elements. This process will be demonstrated in more detail in a later experiment.

For both the voltmeter and the ammeter, always start with the higher ranges and work down to the operating level to avoid damaging the instrument. When the VOM and DMM are returned to the stockroom, be sure the VOM is on the highest voltage scale and the DMM is in the off position.

The voltage across and the current through a resistor can be used to determine its resistance using Ohm's law in the following form:

$$R = \frac{V}{I}$$ **(3.1)**

The magnitude of R will be determined by the units of measure for V and I.

PROCEDURE

The purpose of this laboratory exercise is to acquaint you with the equipment, so *do not rush*. If you are a member of a squad, don't let one individual make all the measurements. You must become comfortable with the instruments if you expect to perform your future job function in a professional manner. Read the instruments carefully. The more accurate a reading, the more accurate the results obtained. One final word of caution. For obvious reasons, *do not make network changes with the power on!* If you have any questions about the procedure, be sure to contact your instructor.

Part 1   Setting the Output Voltage of a dc Power Supply with a DMM and VOM

(a) Connect the DMM to the dc power supply as shown in Fig. 3.7.

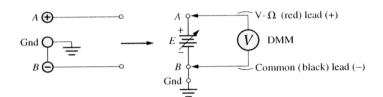

FIG. 3.7

Using the DMM, set the power supply to the voltage levels E_{AB} appearing in Table 3.1. Ignore the meter on the supply when setting the voltage levels. For each setting, choose the scale of the DMM that will result in the highest degree of accuracy. Once a particular level is set, remove the DMM and measure the same voltage with the VOM using the scale of the meter that results in the most accurate reading. Do not be influenced by the level set with the DMM. Simply remove the DMM from the supply when the voltage E_{AB} is set and measure the terminal voltage with the VOM.

TABLE 3.1

E_{AB} (DMM)	VOM	% Difference
1 V		
4 V		
5.5 V		
8.25 V		
9.6 V		
12.1 V		
16.4 V		
18.75 V		

Calculate the magnitude of the percent difference between the DMM setting and the VOM reading using the following formula and complete Table 3.1.

$$\% \text{ Difference} = \left| \frac{\text{DMM} - \text{VOM}}{\text{DMM}} \right| \times 100\% \qquad (3.2)$$

Is the magnitude of the percent difference for each level sufficiently small to verify the fact that the reading of one meter will be very close to the other even though one is analog and the other digital?

(b) This part of the experiment will provide some additional practice in the use of the DMM and VOM. The supply voltage will now be set by the VOM and the setting checked by the DMM. Set the voltage levels indicated in Table 3.2 with the VOM and then measure the set level with the DMM. For each setting, calculate the magnitude of the percent difference using Eq. (3.3) and complete Table 3.2.

$$\% \text{ Difference} = \left| \frac{\text{VOM} - \text{DMM}}{\text{VOM}} \right| \times 100\% \qquad\qquad (3.3)$$

TABLE 3.2

E_{AB} (VOM)	DMM	% Difference
3 V		
4.2 V		
6.75 V		
10.0 V		
14.15 V		

How do the magnitudes of the percent differences of Table 3.2 compare to those of Table 3.1? Can you make any general conclusions based on the results?

(c) We will now investigate the effect of reversing the leads of the meter when measuring a voltage. Using the setup of Fig. 3.7, reset the voltage E_{AB} to 5 V using the DMM. Then disconnect the DMM and connect the red, or V-Ω, lead to the B terminal and the black, or COM, lead to the A terminal. What is the effect on the reading?

Repeat the previous reading using the VOM and the connections just described. What is the effect on the reading?

(d) Based on the results of parts 1(a)–(c), which meter do you prefer to use? Does one appear more accurate? What are the relative advantages of one compared to the other? Answer each question in sentence form.

Part 2   Ohm's Law (Determining *I*)

In this section, the current of a dc series circuit will be determined by a direct measurement and using Ohm's law. In practice, most current levels are determined using Ohm's law and a measured voltage level to avoid having to break the circuit to insert the ammeter. However, one should be aware of the procedure associated with using an ammeter, and one should feel confident that the measured value and that calculated using Ohm's law are very close in magnitude.

(a) Construct the circuit of Fig. 3.8 using the DMM as a milliammeter. Be sure the milliammeter is connected so that conventional current enters the red (positive) terminal of the meter and leaves the black (negative) terminal to ensure a positive reading. Insert the measured value of R in Fig. 3.8 and Table 3.3. Initially set the DMM on the high milliammeter scale. For most DMMs, the red, or positive, lead must be moved from the V-Ω connection to the A terminal of the meter. The COM connection remains the same.

Adjust the power supply until $V_R = 2$ V (the voltage across the resistor, not the supply voltage) using the VOM to monitor V_R. Be sure the red (positive) lead is connected to the point of higher potential (the terminal that conventional current enters) and the black (negative) lead is connected to the point of lower potential (the terminal that conventional current leaves). You may find that searching for the best scale for the milliammeter will affect the voltage across V_R, since changing ammeter scales will change the internal resistance of the milliammeter. Find a scale that provides a reading of good accuracy with V_R set at the required 2 V.

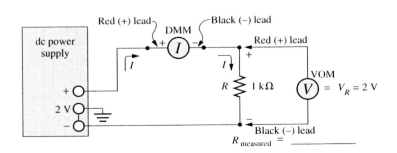

FIG. 3.8

In Table 3.3, record the measured value of I_R from the DMM. Then calculate the level of I_R using Ohm's law and the measured resistor value and record in Table 3.3 (using mA as the unit of measurement for I_R). Finally, determine the magnitude of the percent difference from the following equation and complete the line for $V_R = 2$ V in Table 3.3.

$$\% \text{ Difference} = \left| \frac{I_R \, (\text{DMM}) - I_R \, (\text{Ohm's law})}{I_R \, (\text{DMM})} \right| \times 100\% \qquad (3.4)$$

Repeat this procedure for the other levels of V_R in Table 3.3. Note that when $V_R = 0$ V, $I_R = 0$ mA and percent difference = 0%.

TABLE 3.3 $R = 1$ kΩ $\qquad R_{\text{measured}} = \underline{\hspace{3cm}}$

V_R (VOM)	I_R (DMM) mA	$I_R = V_R/R_{\text{meas.}}$ mA	% Difference
0 V	0 mA	0 mA	0%
2 V			
4 V			
6 V			
8 V			
10 V			

Comment on the level of percent difference in Table 3.3. Are the percent differences sufficiently small to establish firmly the fact that the current determined by Ohm's law will be very close (if not equal) to that measured directly? Answer the question in sentence form.

Part 3   Plotting Ohm's Law

(a) Using the data (measured values) of Table 3.3, plot I (DMM) versus V_R (VOM) on Graph 3.1. Clearly indicate each data point on the graph. Also label the curve as $R = 1$ kΩ.

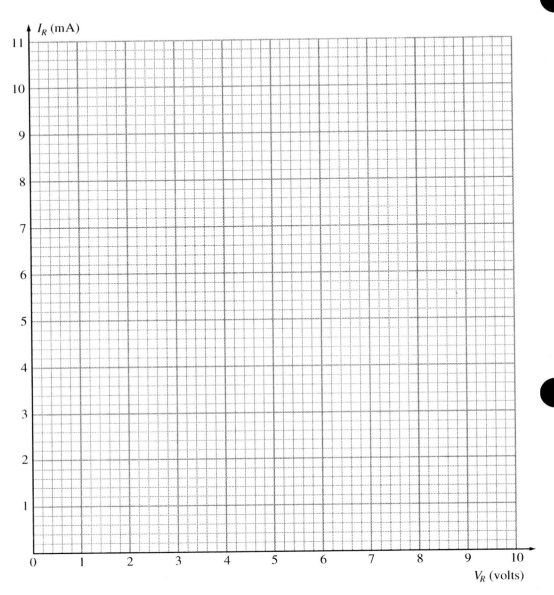

GRAPH 3.1

(b) Once the curve of part (a) is drawn, the level of resistance can be determined at any level of voltage or current.

For instance, at $I_R = 5.6$ mA, draw a horizontal line from the vertical axis to the curve. Then draw a line down from the intersection to the horizontal voltage axis. Record the level of V_R in Table 3.4. Calculate the resistance using Ohm's law and insert in Table 3.4.

Calculation:

Using a similar procedure, determine the level of V_R corresponding to $I_R = 1.2$ mA. Determine the value of R using Ohm's law and compare with the level at $I_R = 5.6$ mA. Record both results in Table 3.4.

Calculation:

To continue, determine the level of I_R corresponding to $V_R = 8.3$ V and calculate the resulting resistance level. Again, record both results in Table 3.4.

Calculation:

TABLE 3.4

I_R (mA)	V_R (V)	R (Ω)
5.6		
1.2		
	8.3	

(c) The resistance level also can be determined from the equation

$$R = \frac{\Delta V_R}{\Delta I_R} \qquad (3.5)$$

where ΔV is the change in V due to a change in current ΔI (or vice versa), as demonstrated in Fig. 3.9.

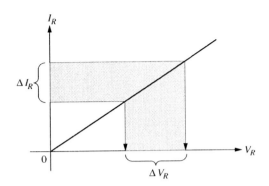

FIG. 3.9

  For instance, if we choose $\Delta I_R = 6$ mA $- 2$ mA $= 4$ mA for the 1-kΩ resistor of Graph 3.1, we can determine the resulting ΔV_R and apply Eq. (3.4). That is, draw a horizontal line from $I_R = 2$ mA and 6 mA on the vertical axis to the curve and then drop lines down to the horizontal axis to determine the corresponding values of V_R. Find the resulting change in V_R and apply Eq. (3.5).

  Determine ΔV_R for $\Delta I_R = 6$ mA $- 2$ mA $= 4$ mA for the 1-kΩ resistor of Graph 3.1. and record in Table 3.5.

  Determine R using Eq. (3.5) and record in Table 3.5.

Calculation:

  Determine ΔI_R for $\Delta V_R = 4.6$ V $- 3.2$ V $= 1.4$ V for the 1-kΩ resistor of Graph 3.1 and record in Table 3.5.

  Determine R using Eq. (3.5) and record in Table 3.5.

Calculation:

TABLE 3.5

ΔV_R (V)	ΔI_R (mA)	R (Ω)
	4	
1.4		

(d) The slope of a curve is related to the resistance by

$$\boxed{\text{Slope} = m = \frac{\Delta y}{\Delta x} = \frac{\Delta I_R}{\Delta V_R} = \frac{1}{R}} \qquad \text{(siemens S)} \qquad \textbf{(3.6)}$$

revealing that the smaller the resistance, the steeper the slope, or the more the resistance, the less the slope.

Determine the slope for the 1-kΩ resistor in mS using the measured resistor value from Fig. 3.8 and record in Table 3.6.

Calculation:

TABLE 3.6

R	m
1 kΩ	
3.3 kΩ	

Part 4   Plotting R = 3.3 kΩ

(a) Reconstruct the circuit of Fig. 3.8 using R = 3.3 kΩ. Insert the measured value of R in Table 3.5 and use this value for all the calculations.

Using the procedure described in part 2, complete Table 3.6.

TABLE 3.7 R = 3.3 kΩ R_{measured} = _____

V_R (VOM)	I_R (DMM) mA	$I = V_R/R_{\text{meas.}}$ mA	% Difference
0 V	0 mA	0 mA	0%
2 V			
4 V			
6 V			
8 V			
10 V			

(b) Using the data of Table 3.7, plot I_R (DMM) versus V_R (VOM) on Graph 3.2. Clearly indicate each data point on the graph. Also label the curve as R = 3.3 kΩ.

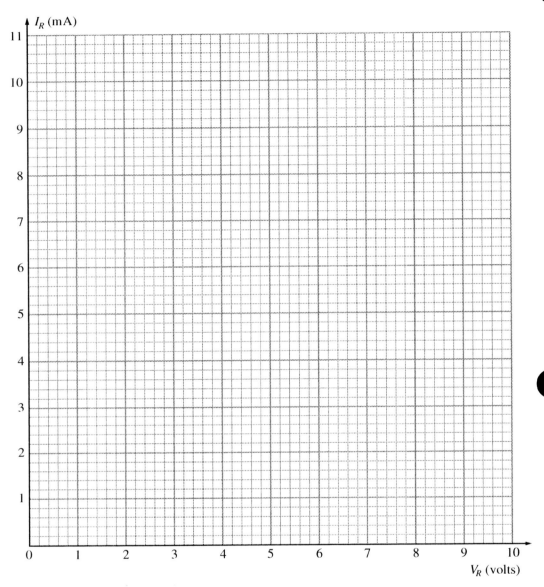

GRAPH 3.2

  (c) Determine the level of V_R corresponding to $I_R = 2.4$ mA and calculate the resistance using Eq. (3.1.). Record both results in Table 3.8.

Calculation:

(d) Determine the ΔV_R corresponding to $\Delta I_R = 2.2 \text{ mA} - 1.4 \text{ mA} = 0.8 \text{ mA}$ and calculate R using Eq. 2.4. Record both results in Table 3.8.

Calculation:

TABLE 3.8

$I_R = 2.4 \text{ mA}$	$\Delta I_R = 0.8 \text{ mA}$
$V_R =$	$\Delta V_R =$
$R =$	$R =$

(e) Determine the slope of the 3.3-kΩ resistor using the measured value (from Table 3.7) and Eq. 3.6 and record in Table 3.6.

Calculation:

How does the magnitude of the slope compare to the magnitude determined for the 1-kΩ resistor? Is the following conclusion verified: The larger the resistance, the less the slope?

EXERCISES

1. Plot the linear curve for 100-Ω and 10-kΩ resistors on Graph 3.3.

2. As the resistance increases, does the slope defined by $m = \Delta I_R / \Delta V_R$ increase or decrease?

3. Under ideal conditions, is the plot of I_R versus V_R for a fixed resistor always a straight line that intersects $I_R = 0$ A and $V_R = 0$ V?

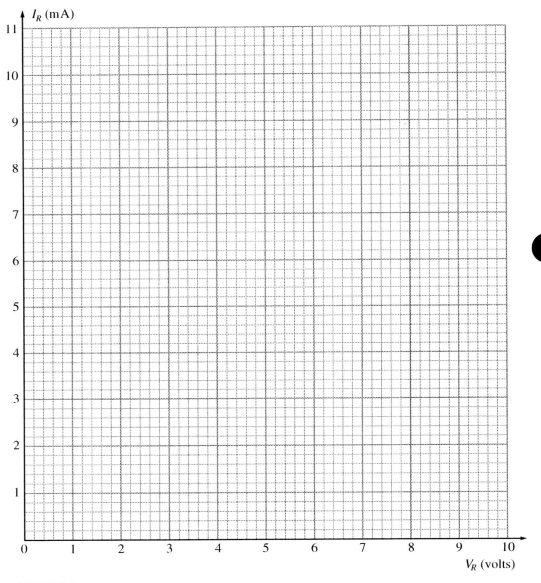

GRAPH 3.3

4. Which two terminals of a dc power supply must always be connected to obtain the desired voltage?

EXPERIMENT dc

4

Series Resistance

OBJECTIVES

1. **Determine the total resistance of a series dc circuit using an ohmmeter or an application of Ohm's law.**
2. **Calculate the total resistance of a series dc circuit and note the effect of the relative magnitude of each series resistor on the total resistance.**
3. **Learn to identify which resistors of a network are in series.**
4. **Develop additional confidence in the use of the DMM ohmmeter.**

EQUIPMENT REQUIRED

Resistors

1—220-Ω, 330-Ω, 1-kΩ, 100-kΩ (1/4-W)

3—100-Ω (1/4-W)

Instruments

1—DMM

1—dc power supply

EQUIPMENT ISSUED

TABLE 4.0

Item	Manufacturer and Model No.	Laboratory Serial No.
DMM		
Power supply		

RÉSUMÉ OF THEORY

The total resistance R_T of a series circuit is the sum of the individual resistances. See Fig. 4.1.

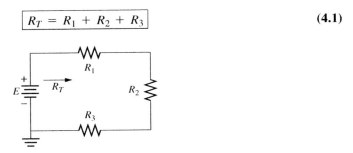

$$R_T = R_1 + R_2 + R_3 \qquad \text{(4.1)}$$

FIG. 4.1

For equal series resistors, the total resistance is equal to the resistance of one resistor times the number in series; that is,

$$R_T = NR \qquad \text{(4.2)}$$

For any system, such as that appearing in Fig. 4.2, the total resistance can be determined by the following form of Ohm's law:

$$R_T = \frac{E}{I} \qquad \text{(4.3)}$$

That is, for an applied voltage E the current is measured and Eq. (4.3) is applied to determine the resistance of the network.

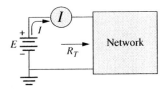

FIG. 4.2

PROCEDURE

Part 1   Two Series Resistors

(a) Construct the circuit of Fig. 4.3.

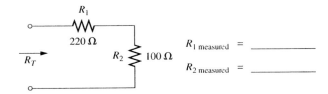

$R_{1\ \text{measured}}$ = _____

$R_{2\ \text{measured}}$ = _____

FIG. 4.3

In the space provided in Fig. 4.3, insert the measured value of each resistor as determined using an ohmmeter.

(b) Calculate the total resistance using the measured resistor values and record in Table 4.1 Show all work! Apply units to all results.

Calculation:

(c) Measure the total resistance using the ohmmeter section of your DMM, and record in Table 4.1.

How do the measured and calculated values compare? Calculate the magnitude of the percent difference from

$$\%\text{Difference} = \left| \frac{\text{Measured} - \text{Calculated}}{\text{Calculated}} \right| \times 100\% \qquad \textbf{(4.4)}$$

and record in Table 4.1.

Calculation:

% Difference = _____

(d) Set the supply to 8 V using the DMM. Then turn off the supply and construct the circuit of Fig. 4.4. Use the same DMM as a milliammeter (be sure to set up the DMM as a milliammeter by choosing the A (ampere) terminal for the positive lead and putting the meter in series with the supply, with conventional current entering the positive terminal and leaving the negative, or COM, terminal). We are unaware of the current levels, so set the ammeter to the highest available scale. Once the circuit is constructed and carefully checked, turn on the supply and record the ammeter reading in Table 4.1. Be sure to use the scale of the ammeter that provides the most accurate reading.

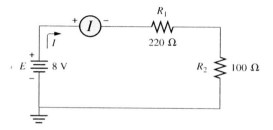

FIG. 4.4

(e) Using the supply voltage and ammeter reading of part 1(d), calculate the total resistance using Eq. (4.3) and record in Table 4.1.

Calculation:

Calculate the percent difference between this part and part 1(c) and insert in Table 4.1.

Calculation:

TABLE 4.1

R_T (calculated)	R_T (ohmmeter)	% Difference	I (measured)	R_T (Ohm's law)	% Difference

Part 2   Three Series Resistors

(a) Construct the circuit of Fig. 4.5. Insert the measured value of each resistor in the space provided.

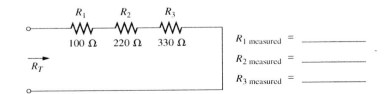

FIG. 4.5

(b) Calculate the total resistance using the measured resistor values and record in Table 4.2. Show all work! Apply units to all results.

Calculation:

(c) Measure the total resistance using the ohmmeter section of your DMM and record in Table 4.2.

How do the calculated and measured values compare? Calculate the magnitude of the percent difference using Eq. (4.4) and record in Table 4.2.

Calculation:

(d) Apply 8 V and an ammeter in the same manner described in part 1(d) and record the current in Table 4.2.

(e) Using the supply voltage and ammeter reading of part 2(d), calculate the total resistance using Eq. (4.3) and record in Table 4.2.

Calculation:

Calculate the percent difference between this part and part 2(c) and record in Table 4.2.

Calculation:

TABLE 4.2

R_T calculated	R_T (ohmmeter)	% Difference	I (measured)	R_T (Ohm's law)	% Difference

Part 3   A Composite

(a) Construct the circuit of Fig. 4.6. Insert the measured value of each resistor in the space provided.

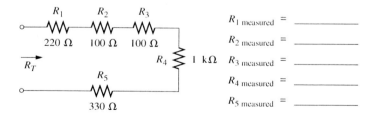

FIG. 4.6

(b) Calculate the total resistance using the measured values and record in Table 4.3. Show all work!

Calculation:

(c) Measure the total resistance using the DMM and find the magnitude of the percent difference between the calculated and measured values. Record both results in Table 4.3.

Calculation:

(d) Apply 8 V as demonstrated in parts 1 and 2 and measure the source current I. Then calculate the total resistance using Eq. (4.3) and compare with the measured value of part 3(c). Record the results in Table 4.3.

Calculation:

TABLE 4.3

R_T(calculated)	R_T(ohmmeter)	% Difference	I (measured)	R_T(Ohm's law)	% Difference

Part 4   Equal Series Resistors

(a) Construct the circuit of Fig. 4.7. Insert the measured value of each resistor in the space provided.

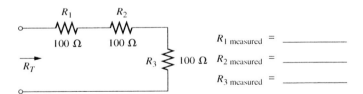

FIG. 4.7

(b) Using the nominal resistor values (all 100 Ω), calculate the total resistance using Eq. (4.2) and record in Table 4.4.

Calculation:

(**c**) Measure the total resistance using the ohmmeter section of your DMM and record in Table 4.4.

How do the measured and calculated values compare? Calculate the magnitude of the percent difference using Eq. (4.4) and insert in Table 4.4.

Calculation:

Is the percent difference sufficiently small to permit the use of Eq. (4.2) and nominal values on an approximate basis? Answer the question in sentence form.

(**d**) Calculate the total resistance using the measured values of each resistor and insert in Table 4.4.

Calculation:

Calculate the percent difference between this result and the measured value of part 4(c) and insert in Table 4.4.

Calculation:

TABLE 4.4

R_T (nominal)	R_T (DMM)	% Difference	R_T (measured)	% Difference

Part 5   Different Levels of Resistance

(a) Construct the circuit of Fig. 4.8. Insert the measured value of each resistor in the space provided.

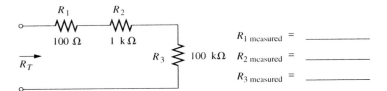

FIG. 4.8

R_1 measured = _____

R_2 measured = _____

R_3 measured = _____

(b) Using the measured values, calculate the total resistance of the circuit and record in Table 4.5. Show all work!

Calculation:

(c) Using the ohmmeter section of the DMM, measure the total resistance and record in Table 4.5.

(d) Calculate the total resistance using measured values if the resistance R_1 is ignored because it is so small in magnitude compared to the 100-kΩ resistor. Record the result in Table 4.5.

Calculation:

Using the ohmmeter section of the DMM, measure the total resistance with R_1 removed and record in Table 4.5.

What is the magnitude of the percent difference between this measured R_T and the measured value of part 5(c)? Record the result in Table 4.5.

Calculation:

(e) Calculate the total resistance using the measured values if both the resistances R_1 and R_2 are ignored. Insert the result in Table 4.5.

Calculation:

Using the ohmmeter section of the DMM, measure the total resistance with R_1 and R_2 removed and record in Table 4.5.

What is the magnitude of the percent difference between this measured R_T and the measured value of part 5(c)? Record the result in Table 4.5.

Calculation:

(**f**) Noting the percent differences of parts 5(d) and 5(e), is it a good first approximation to assume the total resistance of a series circuit is equal to the largest resistor if its magnitude is much greater than the other resistors of the circuit?

(**g**) Apply 10 V to the network of Fig. 4.8 and measure the resulting current I, as demonstrated in previous parts. Record the level of I in Table 4.5. Calculate the total resistance of the circuit and insert the result in Table 4.5. Be aware that the increased level of R_T for this configuration will result in a current in the microampere range. Choose the scale that provides the highest degree of accuracy.

Calculation:

(**h**) Determine the magnitude of the percent difference between the measured value of part 5(g) and the calculated value of part 5(b) and insert in Table 4.5.

Calculation:

TABLE 4.5

R_T (calculated)	R_T (measured)	R_T' (R_1 ignored)	R_T' (measured)	% Difference

R_T'' (R_1, R_2 ignored)	R_T'' (measured)	% Difference	I (measured)	R_T (calculated)	% Difference

Part 6

(a) Construct the circuit of Fig. 4.9. Note the absence of a connection between the lower parts of R_2 and R_3. Insert the measured value of each resistor in the space provided.

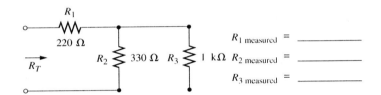

FIG. 4.9

(b) Calculate the total resistance of the configuration using the measured resistor values and record in Table 4.6. Show all work.

Calculation:

(c) Measure the total resistance of the circuit of Fig. 4.9 using the DMM and record in Table 4.6.

(d) Calculate the magnitude of the percent difference between the measured value of part 6(c) and the calculated level of part 6(b) and record in Table 4.6.

Calculation:

(e) If the percent difference is less than a few percent, your theoretical calculations of part (b) were probably correct. What was the impact of the 1-kΩ resistor on the total resistance?

TABLE 4.6

R_T (calculated)	R_T (measured)	% Difference

Part 7

(a) Construct the network of Fig. 4.10. Note that a lead must be placed across the 330-Ω resistor. Insert the measured value of each resistor in the space provided.

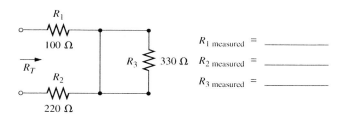

FIG. 4.10

(b) Calculate the total resistance using the measured resistor values and insert in Table 4.7. Show all work!

Calculation:

(c) Measure the total resistance of the network of Fig. 4.10 using the DMM and record in Table 4.7.

(d) Calculate the magnitude of the percent difference between the measured value of part 7(c) and the calculated value of part 7(b) and record in Table 4.7.

Calculation:

(e) If the percent difference is less than a few percent, your theoretical calculations of part 7(b) were probably correct. How did you determine which resistors of Fig. 4.10 were in series?

TABLE 4.7

R_T (calculated)	R_T (measured)	% Difference

EXERCISES

1. Which resistors of Fig. 4.11 are in series?

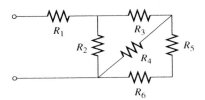

FIG. 4.11

Solution: _____

2. Given the voltage levels of Fig. 4.12, determine the total resistance, R_T, of the system and the resistance of the container, R_T. Show all calculations.

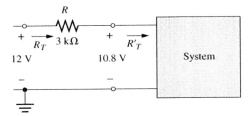

FIG. 4.12

$R_T =$ _____

$R'_T =$ _____

3. Given the information appearing in Fig. 4.13, determine the resistance R_2.

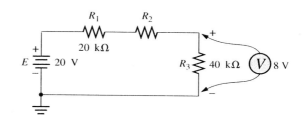

FIG. 4.13

$R_2 = $ _____

Series dc Circuits

OBJECTIVES

1. Correctly measure the voltages and current of a series dc circuit.
2. Verify Kirchhoff's voltage law.
3. Test the application of the voltage divider rule.
4. Become increasingly familiar with the use of the DMM and VOM meters.

EQUIPMENT REQUIRED

Resistors

1—100-Ω, 220-Ω, 330-Ω, 470-Ω, 680-Ω, 1-kΩ, 1-MΩ (1/4-W)

Instruments

1—DMM

1—dc power supply

EQUIPMENT ISSUED

TABLE 5.0

Item	Manufacturer and Model No.	Laboratory Serial No.
DMM		
Power supply		

RÉSUMÉ OF THEORY

In a series circuit, the current is the same through all the circuit elements, as shown by the current I in Fig. 5.1.

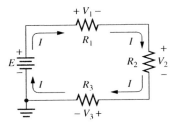

FIG. 5.1

By Ohm's law, the current I is $I = E/R_T$, where E is the impressed voltage and R_T the total resistance. Applying Kirchhoff's voltage law around the closed loop of Fig. 5.1, we find

$$E = V_1 + V_2 + V_3 \qquad (5.1)$$

with $V_1 = IR_1$, $V_2 = IR_2$, and $V_3 = IR_3$.

The voltage divider rule states that the voltage across an element or across a series combination of elements in a series circuit is equal to the resistance of the element or series combination of elements divided by the total resistance of the series circuit and multiplied by the total impressed voltage. For the elements of Fig. 5.1,

$$V_1 = \frac{R_1 E}{R_1} = \frac{R_1 E}{R_1 + R_2 + R_3}, \quad V_2 = \frac{R_2 E}{R_T}, \quad V_3 = \frac{R_3 E}{R_T} \qquad (5.2)$$

PROCEDURE

Part 1   Basic Measurements

(a) Construct the circuit of Fig. 5.2.

Record the measured value of each resistor and set $E = 12$ V using the DMM.

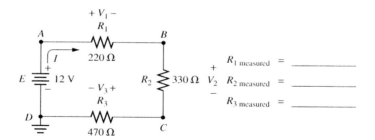

FIG. 5.2

Measure the voltage across the supply and across each resistor with the DMM, making sure to note the polarity of the voltage across each element. Record the voltages in Table 5.1. Use the scale that provides the most accurate reading.

From your readings of the voltages, is Kirchhoff's voltage law satisfied around path $ABCD$? That is, does $E = V_1 + V_2 + V_3$? Answer the question in sentence form.

(b) Now calculate the current through each resistor using Ohm's law and the measured voltage across each resistor and the measured resistor value. Record your results in units of mA in Table 5.1. Show all work.

Calculation:

TABLE 5.1

Voltage (V)		Current (mA)	
E		I_{R_1}	
V_1		I_{R_2}	
V_2		I_{R_3}	
V_3			

How do the levels of I_1, I_2, and I_3 compare? Answer the question in sentence form.

What can you deduce about the current in a series circuit from the preceding calculations?

(c) The current I can be measured by inserting the multimeter (in the ammeter mode) in series with the source, as shown in Fig. 5.3. Be sure the polarity of the meter (defining how the red [+] and black [−] leads are connected) matches that of the voltage across a resistor appearing in the same location as the meter.

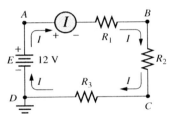

FIG. 5.3

Measure the current at positions A, B, C, and D of Fig. 5.3 and record in Table 5.2.

TABLE 5.2

	I (mA)
A	
B	
C	
D	

Do your results verify the conclusions of part 1(b) regarding the current in a series circuit?

(d) Using the measured resistor values, calculate the total resistance and the current (in mA) for the circuit and record in Table 5.3.

Calculation:

(e) Compute the total resistance of the circuit using the following equation and the current at point A (the source current) from part 1(c) and record in Table 5.3.

$$R_T = \frac{E}{I}$$

(5.3)

How does this value of R_T compare with that obtained in part 1(d)?

(f) Shut down and disconnect the power supply. Then measure the input resistance (R_T) across points A–D using the DMM. Record that value in Table 5.3.

Compare this value with those in parts 1(d) and 1(e).

TABLE 5.3

R_T (calculated)	I (calculated)	R_T (measured)	R_T (ohmmeter)

Part 2   Voltage Divider Rule

Construct the circuit of Fig. 5.4. Insert the measured resistor values in the space provided.

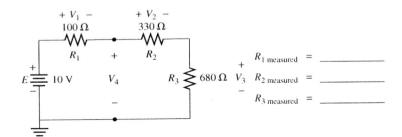

$R_{1\ measured}$ = _____

$R_{2\ measured}$ = _____

$R_{3\ measured}$ = _____

FIG. 5.4

(a) The resistor R_2 is about three times that of R_1. How would you expect the voltages V_2 and V_1 to compare?

Measure V_1 and V_2 and record in Table 5.4. Comment on whether the above assumption was verified.

(b) The resistor R_3 is about two times R_2 and six times R_1. How would you expect the voltage V_3 to compare with the voltages V_2 and V_1?

Measure the voltage V_3 and record in Table 5.4. Comment on whether the above assumption was verified.

(c) Calculate V_3 using the voltage divider rule and measured resistor values. Record in Table 5.4. Show all work!

Calculation:

(d) Determine the magnitude of the percent difference between the measured value of V_3 from part 2(b) and the calculated value of part 2(c) using the following equation: Record in Table 5.4.

$$\% \text{ Difference} = \left| \frac{V_{calc.} - V_{meas.}}{V_{calc.}} \right| \times 100\% \qquad (5.4)$$

Calculation:

TABLE 5.4

V_1 (measured)	V_2 (measured)	V_3 (measured)	V_3 (calculated)	% Difference	V_4 (calculated)	V_4 (measured)	% Difference

(e) Using the voltage divider rule and measured resistor values, calculate the voltage V_4 (the voltage across R_2 and R_3) and record in Table 5.4. Show all work.

Calculation:

(f) Measure V_4 and determine the magnitude of the percent difference with the calculated value of part 2(e). Record both quantities in Table 5.4.

Calculation:

(g) Construct the network of Fig. 5.5.

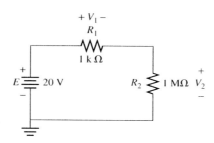

FIG. 5.5

(h) By simply noting the relative magnitude of the resistors R_1 and R_2, how would you expect the applied voltage to divide between the series resistive elements? How would you expect the voltage levels V_1 and V_2 to compare to the applied voltage?

(i) Measure the voltages V_1 (in mV) and V_2 (in V) and determine whether your conclusions of part 3(e) were correct.

$V_1 =$ _____ $V_2 =$ _____

Part 3   Open Circuits

(a) Construct the network of Fig. 5.6. Insert the measured value of each resistor in the space provided.

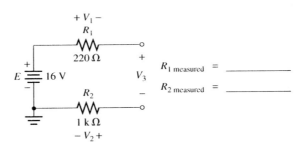

FIG. 5.6

(b) Calculate the voltages V_1, V_2, and V_3 and record in Table 5.5. Explain your results.

Calculation:

(c) Measure voltages V_1, V_2, and V_3 with the DMM and record in Table 5.5.

Do the measured values validate the conclusion of part 3(b)?

TABLE 5.5

	Calculated	Measured
V_1		
V_2		
V_3		

EXERCISES

1. Given the voltage V_1 in Fig. 5.7, determine the resistor R_1 using the voltage divider rule.

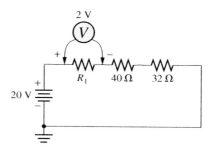

FIG. 5.7

$R_1 = $ _____

2. Determine V_1 for the configuration of Fig. 5.8.

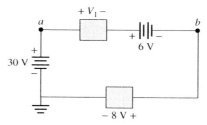

FIG. 5.8

$V_1 =$ _____

3. For the system of Fig. 5.8, determine the DMM reading if the red, or positive, lead is connected to point a and the black, or negative, lead is connected to point b.

$V_{ab} =$ _____

4. For the series circuit of Fig. 5.2 record the measure values V_1, V_2, V_3 and the current I in Table 5.6 (from Table 5.1).

   Then, using PSpice or Multisim, determine the same voltages and current and enter the results in Table 5.6. Use measured values for the resistors.

TABLE 5.6

	Table 5.1	PSpice or Multisim
V_1		
V_2		
V_3		
I		

How do the results compare? Explain any differences. Attach the appropriate printouts.

Name:_____

Date:_____

Course and Section:_____

Instructor:_____

EXPERIMENT dc **6**

Parallel Resistance

OBJECTIVES

1. Determine the total resistance of a parallel network using an ohmmeter or an application of Ohm's law.
2. Note the effect of the relative magnitude of each parallel resistor on the total resistance.
3. Learn to identify which resistors of a network are in parallel.

EQUIPMENT REQUIRED

Resistors

1—100-Ω, 1-kΩ, 1.2-kΩ, 2.2-kΩ, 100-kΩ (1/4-W)

3—3.3-kΩ (1/4-W)

Instruments

1—DMM

1—dc power supply

EQUIPMENT ISSUED

TABLE 6.0

Item	Manufacturer and Model No.	Laboratory Serial No.
DMM		
Power supply		

RÉSUMÉ OF THEORY

In a parallel circuit (Fig. 6.1), the total or equivalent resistance (R_T) is given by

$$\frac{1}{R_T} = \frac{1}{R_1} + \frac{1}{R_2} + \frac{1}{R_3} + \cdots \qquad (6.1)$$

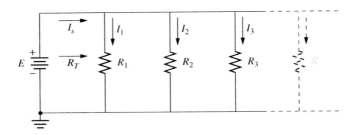

FIG. 6.1

If there are only two resistors in parallel, it is more convenient to use

$$R_T = \frac{R_1 R_2}{R_1 + R_2} \qquad (6.2)$$

For equal parallel resistors, the total resistance is equal to the resistance of one resistor divided by the number (N) of resistors in parallel; that is,

$$R_T = \frac{R}{N} \qquad (6.3)$$

For an unknown system, the same technique applied for series dc circuits can be applied to parallel dc circuits. In Fig. 6.2, the total resistance can be determined from

$$R_T = \frac{E}{I} \qquad (6.4)$$

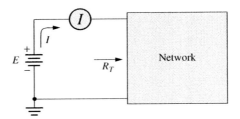

FIG. 6.2

For parallel resistors, the total resistance is always less than the value of the smallest parallel resistor. In fact, the total resistance is very close to the value of one resistor if that resistance is much less than the others.

PROCEDURE

Part 1 Two Parallel Resistors

(a) Construct the circuit of Fig. 6.3 and record the measured values of R_1 and R_2 in the space provided.

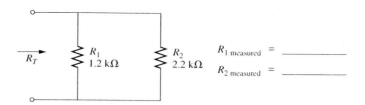

R_1 measured = _____

R_2 measured = _____

FIG. 6.3

(b) Using the measured resistor values, calculate the total resistance using Eq. (6.1) and record in Table 6.1. Show all work. Apply units to the result.

Calculation:

(c) Using the ohmmeter section of your DMM, measure the total resistance of the network of Fig. 6.3 and record in Table 6.1.

(d) How do the results of parts 1(b) and 1(c) compare? Calculate the magnitude of the percent difference from

$$\% \text{ Difference} = \left| \frac{\text{Calculated} - \text{Measured}}{\text{Calculated}} \right| \times 100\% \qquad (6.5)$$

and record in Table 6.1.

Calculation:

TABLE 6.1

R_T (Eq. 6.1)	R (ohmmeter)	% Difference	R_T (Eq. 6.2)	I	R (Ohm's law)

(e) Calculate the total resistance using measured values and Eq. (6.2) and record in Table 6.1. How do the results compare with part 1(b)?

Calculation:

(f) Set the supply to 10 V using the DMM. Then turn off the supply and construct the network of Fig. 6.4. Once constructed and carefully checked, turn on the supply and record the ammeter reading in Table 6.1.

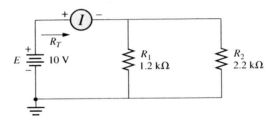

FIG. 6.4

(g) Using the supply voltage and the milliammeter reading of part 1(f), calculate the total resistance using Eq. (6.4) and record in Table 6.1. Ignore the voltage drop across the ammeter.

Calculation:

How do the results of this part and part 1(c) compare?

(h) The Résumé of Theory states that the total resistance of a parallel network will always be less than the smallest resistor. Has this conclusion been verified by the experimental work of this part of the experiment?

Part 2 Three Parallel Resistors

(a) Construct the network of Fig. 6.5.

Insert the measured values of the resistors in the spaces provided.

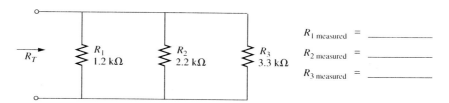

$R_{1 \text{ measured}}$ = _____

$R_{2 \text{ measured}}$ = _____

$R_{3 \text{ measured}}$ = _____

FIG. 6.5

(b) Calculate the total resistance using the measured resistor values and record in Table 6.2. Show all work!

Calculation:

(c) Measure the total resistance using the DMM and calculate the magnitude of the percent difference. Record both results in Table 6.2.

Calculation:

(d) Using the procedure introduced in part 1(f), set the supply voltage to 10 V and measure the source current and record in Table 6.2.

TABLE 6.2

R_T (calculated)	R_T (ohmmeter)	% Difference	I	R_T (Ohm's law)

(e) Using $E = 10$ V and the ammeter reading of part 2(d), calculate the total resistance using Eq. (6.4) and record in Table 6.2.

How do the results of this part and part 2(c) compare?

(f) How does the total resistance compare to the magnitude of the smallest of the parallel resistors? Has the addition of the 3.3-kΩ resistor reduced the total resistance below that of the previous part of the experiment? Will the total resistance of a parallel network always decrease in magnitude with the connection of additional parallel elements? Answer all questions in sentence form.

Part 3   Equal Parallel Resistors

(a) Construct the network of Fig. 6.6.

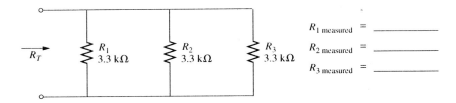

FIG. 6.6

Insert the measured value of each resistor in the space provided.

(b) Assuming all the resistors are their nominal value of 3.3 kΩ, calculate the total resistance using Eq. (6.3) and record in Table 6.3.

Calculation:

(c) Measure the total resistance using the ohmmeter section of your DMM and record in Table 6.3.

How do the calculated and measured values compare? Calculate the magnitude of the percent difference using Eq. (6.5) and record in Table 6.3.

Calculation:

TABLE 6.3

R_T (Nominal)	R_T (ohmmeter)	% Difference	R_T (measured)

Is the percent difference sufficiently small to permit the use of the nominal resistor values as a good first approximation to the actual situation?

(d) Calculate the total resistance using the measured values of each resistor and record in Table 6.3.

Calculation:

How does this calculated value compare with the measured value of part 3(c)?

Part 4   Different Levels of Resistance

(a) Construct the network of Fig. 6.7.

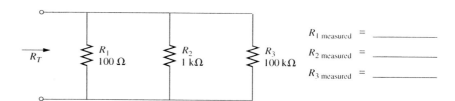

$R_{1 \text{ measured}}$ = _____

$R_{2 \text{ measured}}$ = _____

$R_{3 \text{ measured}}$ = _____

FIG. 6.7

Insert the measured value of each resistor in the space provided.

(b) Using the measured values, calculate the total resistance of the circuit and insert in Table 6.4. Show all work.

Calculation:

(c) Using the ohmmeter section of the DMM, measure the total resistance and insert in Table 6.4.

Compare with the calculated value.

(d) Calculate the total resistance using measured values if the resistance R_3 is ignored and record in Table 6.4.

Calculation:

(e) Calculate the total resistance using the measured values if resistances R_2 and R_3 are ignored. Insert the result in Table 6.4.

Calculation:

TABLE 6.4

R_T (all)	R_T (ohmmeter)	R_T (R_3 ignored)	R_T (R_2, R_3 ignored)

Is there a great deal of difference between the calculated values of parts 4(b), 4(d), and 4(e)? Can you draw any conclusions from the results?

(f) It was noted in the Résumé of Theory that the total resistance of parallel resistors is always less than the smallest parallel resistor. Was this conclusion verified by the results of parts 4(b) and 4(c)?

(g) It was also pointed out in the Résumé of Theory that the total resistance of parallel elements will be very close to the smallest if the smallest is much less than the other parallel elements of the network. Noting the results of parts 4(b) and 4(c), is this conclusion verified?

Part 5   Open Circuits

(a) Construct the network of Fig. 6.8. Note the absence of a connection between the bottom terminals of the last two parallel resistors. Insert the measured value of each resistor in the space provided.

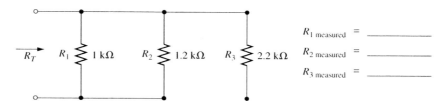

FIG. 6.8

(b) Calculate the total resistance of the configuration using the measured resistor values and insert in Table 6.5. Show all work.

Calculation:

(c) Measure the total resistance of the network of Fig. 6.8 using the DMM and insert in Table 6.5.

(d) Calculate the magnitude of the percent difference between the measured value of part 5(c) and the calculated value of part 5(b) using Eq. (6.5) and record in Table 6.5.

Calculation:

TABLE 6.5

R_T (theory)	R_T (ohmmeter)	% Difference

(e) If the percent difference is less than a few percent, your theoretical calculations of part 5(b) were probably correct. What was the impact of the 2.2-kΩ resistor? Why?

Part 6   Short Circuits

(a) Construct the network of Fig. 6.9. Insert the measured value of each resistor in the space provided. Take special note of the lead connected directly across the resistor R_2.

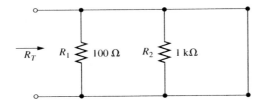

$R_{1\ measured}$ = _____

$R_{2\ measured}$ = _____

FIG. 6.9

(b) Calculate the total resistance of the network of Fig. 6.9 using the measured resistor values and record in Table 6.6.

Calculation:

(c) Measure the total resistance of the network of Fig. 6.9 using the DMM and record in Table 6.6.

(d) Based on the results of parts 6(b) and (c), what was the impact of the lead across R_2 in Fig. 6.9?

TABLE 6.6

R_T (theory)	R_T (ohmmeter)

EXERCISES

1. Which resistors of Fig. 6.10 are in series and which are in parallel?

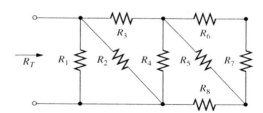

FIG. 6.10

Series resistors: _____
Parallel resistors: _____

2. Using the milliammeter readings provided in Fig. 6.11, determine the total resistance of the network R_T. Assume $R_{internal} = 0 \, \Omega$ for all meters.

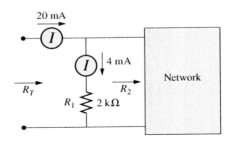

FIG. 6.11

$R_T = $ _____

What is the resistance R_2?

$R_2 = $ _____

Is it necessary to find the resistance R_2 before finding R_T? Explain.

Name: _____

Date: _____

Course and Section: _____

Instructor: _____

Parallel dc Circuits

OBJECTIVES

1. Correctly measure the currents and voltage of a parallel dc network.
2. Verify Kirchhoff's current law.
3. Test the application of the current divider rule.

EQUIPMENT REQUIRED

Resistors

1—1-kΩ, 1.2-kΩ, 3.3-kΩ, 4.7-kΩ, 10-kΩ, 1-MΩ (1/4-W)

2—2.2-kΩ (1/4-W)

Instruments

1—DMM

1—dc power supply

EQUIPMENT ISSUED

TABLE 7.0

Item	Manufacturer and Model No.	Laboratory Serial No.
DMM		
Power supply		

RÉSUMÉ OF THEORY

In a parallel circuit (Fig. 7.1), the voltage across parallel elements is the same. Therefore,

$$V_1 = V_2 = V_3 = E \qquad (7.1)$$

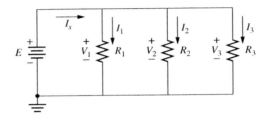

FIG. 7.1

For the network of Fig. 7.1, the currents are related by Kirchhoff's current law,

$$I_s = I_1 + I_2 + I_3 \qquad (7.2)$$

and the current through each resistor is simply determined by Ohm's law:

$$I_1 = \frac{E}{R_1}, \ I_2 = \frac{E}{R_2}, \ I_3 = \frac{E}{R_3}, \ \text{etc.} \qquad (7.3)$$

The current divider rule (CDR) states that the current through one of two parallel branches is equal to the resistance of the *other* branch divided by the sum of the resistances of the two parallel branches and multiplied by the total current entering the two parallel branches. That is, for the networks of Fig. 7.2,

$$I_1 = \frac{R_2 I_s}{R_1 + R_2} \quad \text{and} \quad I_2 = \frac{R_1 I_s}{R_1 + R_2} \qquad (7.4)$$

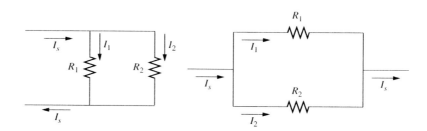

FIG. 7.2

PROCEDURE

Part 1 Basic Measurements

(a) Construct the network of Fig. 7.3. Insert the measured value of each resistor in the space provided.

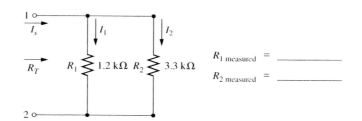

R_1 measured $=$ _____

R_2 measured $=$ _____

FIG. 7.3

(b) Using the measured values, calculate the total resistance of the network and insert in Table 7.1. Show all calculations!

Calculation:

(c) Measure R_T with the ohmmeter section of the DMM and record in Table 7.1. Compare with the result of part 1(b).

TABLE 7.1

R_T (calculated)	R_T (measured)	R_T (Ohm's law)	V_{R_1}	V_{R_2}

(d) How does the total resistance compare with that of the smaller of the two parallel resistors? Will this always be true?

(e) Apply 12 V to points 1–2 of Fig. 7.3 and measure V_{R_1} and V_{R_2} using the DMM.

How do the levels of V_{R_1} and V_{R_2} compare with the applied voltage E? Answer all questions in sentence form.

What general conclusion have you verified regarding the voltage across parallel elements?

(f) Calculate the currents (in mA) through R_1 and R_2 using measured resistor values and Ohm's law and insert in Table 7.2. In addition, determine the source current using Kirchhoff's current law and record in Table 7.2. Show all work!

Calculation:

TABLE 7.2

	Calculated	Measured
I_1		
I_2		
I_s		

(g) Measure the currents I_1, I_2, and I_s using the DMM in the ammeter mode, as shown in Fig. 7.4 and record in Table 7.2. For I_1 and I_2, it is simply a matter of breaking the connections to the top of R_1 and R_2 and inserting the ammeter.

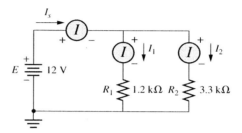

FIG. 7.4

Compare the measured values with the calculated values of part 1(f). Are the differences small enough to verify your theoretical calculations?

(h) Determine the total resistance of the network R_T using Ohm's law and the measured value of I_s and record in Table 7.1. Ignore the voltage drop across the ammeter.

$$R_T = \frac{E}{I_s}$$ (7.5)

Calculation:

How does this level compare with the measured value in Table 7.1?

Part 2   Equal Parallel Resistors

(a) Construct the network of Fig. 7.5. Insert the measured value of each resistor in the space provided.

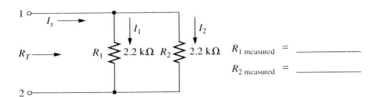

FIG. 7.5

(b) Calculate the total resistance assuming exactly equal resistor values. Use the nominal color-coded value of 2.2 kΩ. Record the result in Table 7.3.

Calculation:

(c) Connect the ohmmeter section of the DMM to points 1–2 and record the reading in Table 7.3. Compare with the result of part 2(b).

TABLE 7.3

R_T (nominal)	R_T (ohmmeter)	R_T (measured)	% Difference

(d) Using the measured resistance levels of R_1 and R_2, calculate the total resistance and record in Table 7.3.

Calculation:

(e) Calculate the percent difference between the calculated value of part 2(b) and the calculated value of part 2(d) using the following equation and record in Table 7.3.

$$\% \text{ Difference} = \left| \frac{R_T(d) - R_T(b)}{R_T(d)} \right| \times 100\% \qquad (7.6)$$

Calculation:

(f) Is the percent difference of part 2(e) sufficiently small to permit the assumptions of part 2(b) on a first approximation basis?

(g) Apply 12 V to points 1–2 of Fig. 7.5 and verify that the voltage across R_1 and R_2 is 12 V. Then measure the currents I_1 and I_2 using the milliammeter section of the DMM and record in Table 7.4.

(h) Do the results of part 2(g) suggest a general conclusion about the division of current through equal parallel elements?

(i) Using the results of part 2(g), calculate the current I_s and record in Table 7.4.

Calculation:

TABLE 7.4

I_1 (measured)	I_2 (measured)	I_s (calculated)	I_s (measured)

(j) Measure the current I_s with $E = 12$ V and record in Table 7.4. Compare to the calculated value shown in Table 7.4. Ignore the voltage drop across the ammeter.

Part 3   Current Division

(a) Construct the network of Fig. 7.6. Insert the measured value of each resistor.

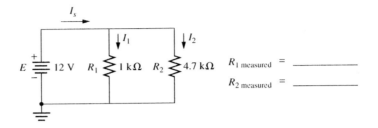

FIG. 7.6

(b) Noting the nominal values of the resistors in Fig. 7.6, how would you expect the currents I_1 and I_2 to be related? That is, how would you expect the magnitude of I_2 to compare to I_1 simply by looking at the magnitude of the resistors?

(c) Measure the currents I_1 and I_2 and record in Table 7.5.

TABLE 7.5

I_1 (measured)	I_2 (measured)	I_s (calculated)	I_3 (calc. 3(c))

(d) Do the results of part 3(c) verify your conclusions of part 3(b)?

(e) Using the measured resistor values, calculate the current I_s and record in Table 7.5.

Calculation:

(f) Calculate the current I_s using the measured values of part 3(c) and record in Table 7.5.

Calculation:

(g) How does the calculated value of part 3(e) compare to the calculated value of part 3(f)?

(h) Construct the network of Fig. 7.7. Insert the measured value of each resistor in the space provided.

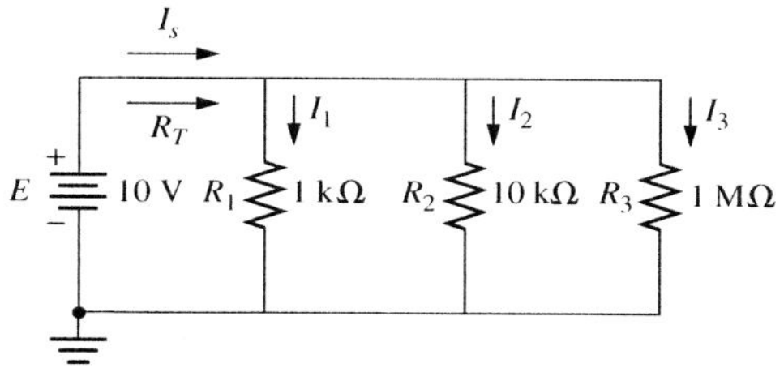

$R_{1\ \text{measured}}$ = ________

$R_{2\ \text{measured}}$ = ________

$R_{3\ \text{measured}}$ = ________

FIG. 7.7

(i) Without making any handwritten calculations, how would you expect the magnitude of the current I_3 to compare to I_1, I_2, or I_s? Why?

(j) Noting the magnitudes of R_1 and R_2, how would you expect the magnitudes of I_1 and I_2 to compare?

(k) Finally, how would you expect the magnitudes of I_s and I_1 to compare?

(l) Measure the currents I_s, I_1, I_2, and I_3 and record in Table 7.6.

TABLE 7.6

I_s	I_1	I_2	I_3	R_T (estimated)	R_T (calculated)

(m) Based on the measurements of part 3(l), are your conclusions of part 3(i) verified?

(n) Based on the measurements of part 3(l), are your conclusions of part 3(j) verified?

(o) Based on the measurements of part 3(l), are your conclusions of part 3(k) verified?

(p) Without making any calculations, what value would you expect for the total resistance R_T of the network of Fig. 7.7? Record in Table 7.6. Explain why you estimated the value you did.

(q) Using the applied voltage E and the measured current I_s, calculate the total resistance R_T and insert in Table 7.6. How does it compare to the estimated value of part 3(p)?

Calculation:

EXERCISES

1. Given the milliammeter readings appearing in Fig. 7.8, determine the resistance R_2. Show all work! Assume $R_{internal} = 0\ \Omega$ for all meters.

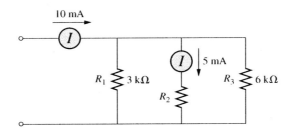

FIG. 7.8

$R_2 = \underline{\hspace{3cm}}$

2. Given the milliammeter readings appearing in Fig. 7.9, determine the resistances R_1 and R_2. Assume $R_{\text{internal}} = 0\ \Omega$ for all meters.

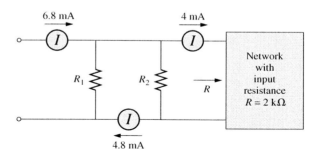

FIG. 7.9

$R_1 = \underline{\hspace{3cm}}, R_2 = \underline{\hspace{3cm}}$

3. For the parallel network of Fig. 7.7 record the values of I_1, I_2, I_3, and I_s in Table 7.7. (from Table 7.6).

Then using PSpice or Multisim, determine the same currents and enter the results in Table 7.7. Use measured values for the resistors.

TABLE 7.7

	Table 7.6	PSpice or Multisim
I_s		
I_1		
I_2		
I_3		

How do the results compare? Try to explain any differences. Attach all appropriate printouts.

Rheostats and Potentiometers

OBJECTIVES

1. Become familiar with the terminal characteristics of a potentiometer.
2. Understand how to use a potentiometer to control potential levels.
3. Learn how to use a potentiometer as a rheostat.
4. Become aware of when a potentiometer is being used as a linear or nonlinear control element.

EQUIPMENT REQUIRED

Resistors

1–470-Ω, 1-kΩ, 10-kΩ, 100-kΩ (1/4-W)

1–1-kΩ linear carbon potentiometer

Instruments

1–DMM

1–dc power supply

EQUIPMENT ISSUED

TABLE 8.0

Item	Manufacturer and Model No.	Laboratory Serial No.
DMM		
Power supply		

RÉSUMÉ OF THEORY

The *potentiometer* (often referred to as a *pot*) is a three-terminal device that is used primarily to control *potential* (voltage) levels. The control can be linear or nonlinear and can extend from a minimum value of 0 V to the maximum available voltage.

The *rheostat* is a two-terminal variable resistance device whose terminal characteristics can also be linear (straight-line) or nonlinear. It is typically used in series with the load to control the voltage and current levels. Its range of control is limited by the maximum resistance of the rheostat.

A three-terminal potentiometer can be used as a two-terminal rheostat if desired. We examine the use of a potentiometer in both modes in this experiment.

PROCEDURE

Part 1   Potentiometer Characteristics

(a) The basic construction and electrical schematic symbol for a potentiometer are provided in Fig. 8.1. Note in Fig. 8.1(b) that there is a continuous resistive element between the external points *A* and *C* that is unaffected by the position of the contact connected to point *B*. As the contact is moved in a clockwise direction around the potentiometer of Fig. 8.1(b), the resistance between points *A* and *B* increases at the same rate as the resistance between points *B* and *C* decreases. Further, note that no matter where the wiper arm is located, the resistance between points *A* and *C* will not change.

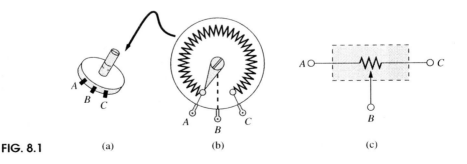

FIG. 8.1   (a)   (b)   (c)

(b) Connect an ohmmeter across the outside two terminals of the 1-kΩ potentiometer, as shown in Fig. 8.2, with terminal *B* (the wiper arm) left open. Turn the control knob as far as it will go in the clockwise (CW) direction and record the resistance in Table 8.1.

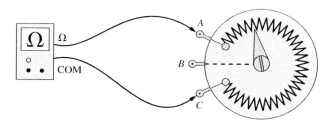

FIG. 8.2

Now turn the control arm in the counterclockwise (CCW) direction as far as it will go and again record the resistance in Table 8.1.

Finally, turn the control arm to any position between the two extremes and record the resulting resistance in Table 8.1.

TABLE 8.1

R (full CW)	R (full CCW)	R (random)

Now review the results of the three measurements and develop a conclusion about the resistance between the two outside terminals of a potentiometer as the wiper arm is moved to any position between the two.

Conclusion:

(c) Now connect the ohmmeter to an outside terminal (either one) and the wiper arm (center connection), as shown in Fig. 8.3, and turn the control knob fully CW. Record the resistance in Table 8.2.

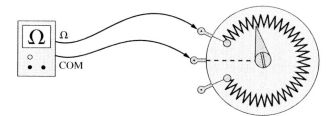

FIG. 8.3

TABLE 8.2

R (full CW)	R (full CCW)	R (full range)	R (CW-other terminal)	R (CCW-other terminal)	R (full range-other terminal)

With the same connections in place, turn the control knob fully CCW and record the resistance in Table 8.2.

Reviewing the results of your last two measurements, determine the full range of the resistance between the center contact and one outside contact and record in Table 8.2.

With the connections still set as in Fig. 8.3, turn the control arm between extremes and note the variation in resistance. Does it appear that turning the control knob through half (an approximation will do) of its full rotation will generate about half the maximum resistance level?

For this linear 1-kΩ potentiometer, does it appear that the resistance between the center contact and one outside contact is directly proportional to the amount we turn the wiper arm?

(d) For this part, maintain the connection to the wiper arm as shown in Fig. 8.3, but move the connection from the outside terminal to the other outside terminal. Turn the control knob to the full CW position again and record the resistance in Table 8.2.

Now turn the knob to the full CCW position and record the total resistance in Table 8.2.

Determine the full range of resistance for these connections and enter in Table 8.2.

Does the full range of resistance match the range using the original connections of Fig. 8.3?

What conclusion can you draw from the above measurements regarding the range of resistance between one outside terminal and the center wiper arm?

Conclusion:

(e) Set the wiper arm in any random position between the full CW and the full CCW position. Then measure the resistance between one outside terminal and the center contact and record in Table 8.3.

Then, without touching the control arm, measure the resistance between the center contact and the other outside terminal and record in Table 8.3.

Finally, add the results of the preceding measurements and record in Table 8.3.

Calculation:

TABLE 8.3

R(random)	R(other contact)	R(Total)

What conclusion can you draw from the exercise just performed?

Conclusion:

Part 2   Potentiometer Control of Potential (Voltage) Levels

(a) Construct the network of Fig. 8.4.

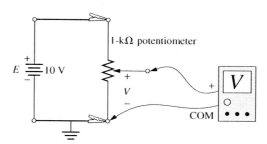

FIG. 8.4

Set the supply voltage to 10 V using the DDM voltmeter. Be sure the two outside terminals of the 1-kΩ potentiometer are connected to the supply as shown. Then connect the voltmeter between the wiper arm and grounded side of the potentiometer.

Turn the control arm and record in Table 8.4 the range of voltage you can control with the potentiometer.

TABLE 8.4

V (range)	R (5V)	R (3.8V)

Adjust the wiper arm until the voltmeter reads 5 V. Then, turn off the power supply, disconnect the potentiometer, and measure the resistance between the two terminals to which the voltmeter was connected and record in Table 8.4.

How does the measured value of resistance compare to the total resistance of the potentiometer?

Reconnect the potentiometer as shown in Fig. 8.4 and adjust the wiper arm until the voltage is 3.8 V. Then disconnect the potentiometer and measure the resistance between the same two terminals to which the voltmeter was connected, recording the result in Table 8.4.

How does this measured value of resistance compare to the total resistance of the potentiometer? Is it 38% of the total resistance to match the 38% of total voltage measured?

What conclusion can you draw from this exercise in regard to the control of voltage levels by a potentiometer?

Conclusion:

(b) The measurements of part (a) were taken without a load connected to the output terminals. The next part of the experiment will demonstrate that the control of the output voltage is very dependent on the load applied. **If the applied load is significantly more than the total resistance of the potentiometer, the operation will be much like the ideal response obtained in part (a).** However, if the applied load has a resistance close to the total resistance of the potentiometer, the control will be severely affected. To demonstrate this effect, first construct the network of Fig. 8.5 and set the output voltage to 5 V. Then connect the 100-kΩ resistor (which is

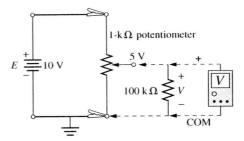

FIG. 8.5

TABLE 8.5

V(100 kΩ)	V(10 kΩ)	V(1 kΩ)	V(470 Ω)

100 times the maximum resistance of the potentiometer) and measure the voltage across the 100-kΩ resistor. Be as accurate as possible. Record the result in Table 8.5.

How much did the voltage drop from its preset value of 5 V?

Is the change in voltage one that you would normally ignore for most applications?

Now, remove the 100-kΩ resistor and reset the voltage to 5 V. However, this time connect a 10-kΩ resistor (10 times the maximum value of the potentiometer) across the preset value and measure the resulting voltage across the 10-kΩ resistor, recording the result in Table 8.5.

How much did the voltage drop from its preset value of 5 V?

Is the change in voltage still small enough that it can be ignored for most applications? If not, explain why.

Now, remove the 10-kΩ resistor and reset the voltage to 5 V. This time, connect a 1-kΩ resistor (equal to the maximum value of the 1-kΩ potentiometer) and measure the resulting voltage across the 1-kΩ resistor. Record in Table 8.5.

How much did the voltage drop from its preset value of 5 V?

Is the change significant and one that cannot be ignored?

As a final example to show the impact of applying a load that is too close in value to the resistance of the potentiometer, connect a 470-Ω resistor across the preset level of 5 V and record the voltage in Table 8.5.

How much did the voltage drop from its preset value of 5 V?

What conclusion can you now draw about the range of resistive loads that can be connected to a potentiometer without affecting the controlled terminal voltage?

Conclusion:

In the problem section, equivalent circuits for some of the situations examined in this part provide an opportunity to investigate further why the voltage dropped as it did with lower and lower applied resistive loads.

Part 3 Using the Potentiometer as a Rheostat

A potentiometer is frequently made into a variable resistor (rheostat) by only using two of its terminals. If we use just the outside two terminals, we now know that the resistance remains fixed at the full resistance level of the potentiometer, no matter how we turned the control arm. However, if we use the wiper arm and only one outside terminal, as shown in Fig. 8.6, the resistance can be varied through the full range of the potentiometer. Because one of the outside terminals is not part of the resulting network, it can be left open, as shown in Fig. 8.6(a), or shorted to the wiper arm, as shown in Fig. 8.6(b). For the configuration of Fig. 8.6(a) the controlling resistance is between points A and B. In Fig. 8.6(b) the direct connection from the wiper arm to terminal C is shorting out the effect of that portion of the potentiometer resistance, leaving the controlling resistance between points A and B.

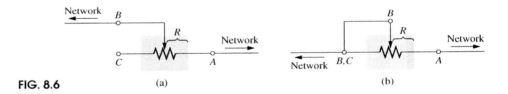

FIG. 8.6   (a)   (b)

We now demonstrate the use of a potentiometer as a rheostat by controlling the voltage to the 470-Ω resistor in Fig. 8.7. Construct the network of Fig. 8.7 and then set the supply voltage to 12 V using a DDM voltmeter. Be sure the milliammeter has been connected properly for an up-scale (positive) reading and the current control knob on the power supply has been turned to its maximum position. Once you have constructed the network and set the supply voltage, connect the voltmeter across the 470-Ω resistor and find the minimum and maximum voltages to which you can set the voltage using the control arm of the rheostat and record in Table 8.6.

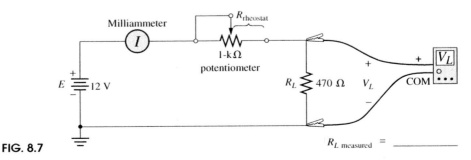

FIG. 8.7

What is the terminal resistance of the rheostat when the voltage across the 470 Ω is at its maximum value? Answer this question from a purely theoretical approach—no measurements.

What is the terminal resistance of the rheostat when the voltage across the 470 Ω is its minimum value? Again, answer this question on a purely theoretical basis.

Set the output voltage to its minimum value and turn off the power supply. Remove the potentiometer and measure the terminal resistance. Record in Table 8.6. Draw the resulting equivalent circuit for Fig. 8.7, replacing the rheostat by a single resistive element. Use this value and the measured value of the 470-Ω resistor and calculate the output voltage using the supply voltage of 12 V and the voltage divider rule. Show all work. Record result in Table 8.6.

Draw an equivalent circuit here:

TABLE 8.6

R_L	V_L (minimum)	V_L (maximum)	R (meas., V_L min.)	V_L (calculated)
470 Ω				
1 kΩ				
10 kΩ				
100 kΩ				

How does this value compare with the measured value for V_L (minimum) in Table 8.6?

What conclusion can you draw for a configuration such as Fig. 8.7 with regard to the range of control of the output voltage it has? Is it sensitive to the applied load or maximum value of the rheostat?

Conclusion:

Now change the 470-Ω resistor to 1 kΩ; proceeding as before, determine the range of control for the output voltage and record the minimum and maximum values of V_L in Table 8.6. Are the results as expected? Explain.

Next, change the 1-kΩ resistor to a 10-kΩ resistor and proceed as described above. Are the results as expected? Explain.

Finally, change the 10-kΩ resistor to a 100-kΩ resistor and proceed as above. Are the results as expected? Explain.

For a rheostat of a given value, what conclusion can you now draw regarding the effect the applied load will have on the range of output voltage?

Conclusion:

Part 4   Current Control with a Two-Point Rheostat

We now demonstrate how a rheostat can control the current through a load and how the rheostat settings can be determined. In addition, we now find, for situations where the applied load is close in magnitude to the maximum value of the rheostat resistance, that the control is nonlinear. The greater the ratio between load and maximum resistance of the rheostat, the more linear the control.

(a) Construct the series network of Fig. 8.8. Then set the supply voltage to 10 V and connect the voltmeter across the 470-Ω load resistor.

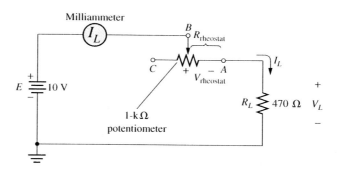

FIG. 8.8

(b) Adjust the rheostat until the voltage across the load is as close to 10 V as possible. Theoretically this would require that the rheostat be set at 0 Ω and the supply be set at exactly 10 V. The resulting current through the load would then be 10 V/470 Ω = 21 mA. However, since your resistor will not be exactly 470 Ω, calculate the current using the measured resistor value and insert this value in the first row, fourth column of Table 8.7. In addition, record the milliammeter reading in the last column.

TABLE 8.7

I_L (Design)	R_{BA} Calculated	V_L Measured	$I_L = V_{L_{meas}}/R_{L_{meas}}$ Calculated	I_L Measured
21 mA	0 Ω	10 V		
18 mA				
14 mA				
10 mA				
8 mA				

(c) We will now calculate the resistance setting of the rheostat to establish a current of 18 mA through the load. The calculations are provided here for the first setting, assuming ideal elements. For each design level you will have to repeat the calculations using the measured resistor value.

$$I_L = 18 \text{ mA} = \frac{V_L}{470 \ \Omega} \Rightarrow V_L = (18 \text{ mA})(470 \ \Omega) = 8.46 \text{ V}$$

$$V_{\text{rheostat}} = E - V_L = 10 \text{ V} - 8.46 \text{ V} = 1.54 \text{ V}$$

$$R_{\text{rheostat}} = \frac{V_{\text{rheostat}}}{I_{\text{rheostat}}} = \frac{1.54 \text{ V}}{18 \text{ mA}}$$

$$= 85.56 \ \Omega$$

For ideal elements, therefore, the rheostat must be set at 85.56 Ω for a load current of 18 mA.

Perform the same calculations using the measured value of the load resistor and insert the result in the second column of Table 8.7.

Now turn off the power supply and remove the 1-kΩ rheostat from Fig. 8.8 and set it for the value just calculated. Return it to the circuit and turn on the power supply again. Measure the load voltage, calculate the load current, and insert the value in Table 8.7. Also record the milliammeter reading.

(d) Repeat this procedure for the other levels of current appearing in Table 8.7. Note that the minimum value is 8 mA, because the theoretical minimum is $I = 10\text{ V}/(1\text{ k}\Omega + 470\text{ }\Omega) = 6.8$ mA, and your resistance may be less than 470 Ω, bringing the minimum current over 7 mA. There is no need to show all your work. Just be very careful with your calculations and measurements. The rheostat resistance should increase and the load voltage should drop for each succeeding level.

(e) When the table is complete, plot the resulting current (I_L calculated) versus the calculated rheostat setting on Graph 8.1. Carefully connect the plot points with a smooth, clear curve. Is the resulting curve linear or nonlinear?

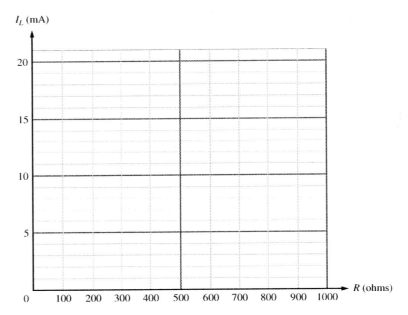

GRAPH 8.1

You will find that the larger the load resistance compared to the maximum rheostat resistance, the more linear (straight-line) the resulting graph. Any load resistance near the maximum rheostat resistance will be quite nonlinear in appearance.

EXERCISES

1. In part 2, the output voltage was set to 5 V before a load of 100 kΩ was applied, resulting in the equivalent network of Fig. 8.9. When preset for 5 V, the wiper arm ideally splits the total resistance of the potentiometer. For the resulting configuration, calculate the load voltage and compare to the measured value of part 2.

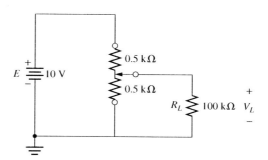

$E \equiv 10\text{ V}$

0.5 kΩ

0.5 kΩ

R_L 100 kΩ V_L + −

FIG. 8.9

V_L (calculated) = _____

V_L (measured from part 2) = _____

2. Repeat the above analysis for the 470 Ω applied load.

V_L (calculated) = _____

V_L (measured from part 2) = _____

3. Current control using a potentiometer can also be established using an arrangement such as the one shown in Fig. 8.10. For the network as it appears, determine the minimum and maximum levels of I_L.

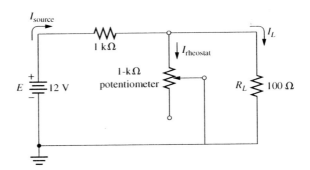

FIG. 8.10

I_L (minimum) = _____

I_L (maximum) = _____

4. Repeat Problem 3 if the load is changed to 1 kΩ.

I_L (minimum) = _____

I_L (maximum) = _____

What relationship between the load resistor level and the maximum potentiometer resistance should be established for good control using the configuration of Fig. 8.10?

EXPERIMENT dc

9

Series-Parallel dc Circuits

OBJECTIVES

1. Test the theoretical analysis of series-parallel networks through direct measurements.
2. Improve skills of identifying series or parallel elements.
3. Measure properly the voltages and currents of a series-parallel network.
4. Practice applying Kirchhoff's voltage and current laws, the current divider rule, and the voltage divider rule.

EQUIPMENT REQUIRED

Resistors

1—1-kΩ, 2.2-kΩ, 3.3-kΩ, 4.7-kΩ (1/4-W)

Instruments

1—DMM

1—dc power supply

EQUIPMENT ISSUED

TABLE 9.0

Item	Manufacturer and Model No.	Laboratory Serial No.
DMM		
Power supply		

RÉSUMÉ OF THEORY

The analysis of series-parallel dc networks requires a firm understanding of the basics of both series and parallel networks. In the series-parallel configuration, you will have to isolate series and parallel configurations and make the necessary combinations for reduction as you work toward the desired unknown quantity.

As a rule, it is best to make a mental sketch of the path you plan to take toward the complete solution before introducing the numerical values. This may result in savings in both time and energy. Always work with the isolated series or parallel combinations in a branch before tying the branches together. For complex networks, a carefully redrawn set of reduced networks may be required to ensure that the unknowns are conserved and that every element has been properly included.

PROCEDURE

Part 1

(a) Construct the series-parallel network of Fig. 9.1. Insert the measured value of each resistor in the space provided.

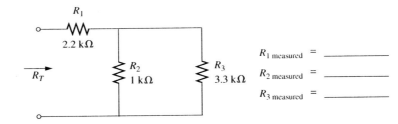

$R_{1 \text{ measured}} =$ _____

$R_{2 \text{ measured}} =$ _____

$R_{3 \text{ measured}} =$ _____

FIG. 9.1

(b) Calculate the total resistance R_T using the measured resistance values and record in Table 9.1.

Calculation:

(c) Use the ohmmeter section of the multimeter to measure R_T and record in Table 9.1.

(d) Determine the magnitude of the percent difference between the calculated and measured values of parts 1(b) and 1(c) using the following equation and record in Table 9.1.

$$\% \text{ Difference } = \left| \frac{\text{Calculated } - \text{ Measured}}{\text{Calculated}} \right| \times 100\% \qquad \textbf{(9.1)}$$

Use Eq. (9.1) for all percent difference calculations in this laboratory experiment.

Calculation:

TABLE 9.1

R_T (calculated)	R_T (measured)	% Difference

(e) If 12 V were applied, as shown in Fig. 9.2, *calculate* the currents I_s, I_1, I_2, and I_3 using the measured resistor values and record in Table 9.2.

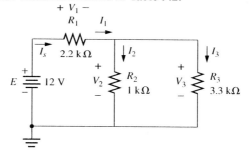

FIG. 9.2

Calculation:

(f) Apply 12 V, *measure* the currents I_1, I_2, and I_3 using the milliammeter section of your multimeter, and record in Table 9.2. Be sure the meter is in series with the resistor through which the current is to be measured. Calculate the magnitude of the percent difference between calculated and measured values using Eq. (9.1) and enter in Table 9.2.

Calculation:

How are the currents I_1 and I_s related? Why?

TABLE 9.2

	Calculated	Measured	%Difference
I_s			
I_1			
I_2			
I_3			

(g) Using the results of part 1(e), *calculate* the voltages V_1, V_2, and V_3 using measured resistor values and record in Table 9.3.

Calculation:

(h) *Measure* the voltages V_1, V_2, and V_3, determine the magnitude of the percent difference between the calculated and measured values, and record the results in Table 9.3.

Calculation:

TABLE 9.3

	Calculated	Measured	%Difference
V_1			
V_2			
V_3			

How are the voltages V_2 and V_3 related? Why?

(i) Referring to Fig. 9.2 and the results in Table 9.3 does $E = V_1 + V_2$, as required by Kirchhoff's voltage law?

Part 2

(a) Construct the series-parallel network of Fig. 9.3. Insert the measured value of each resistor.

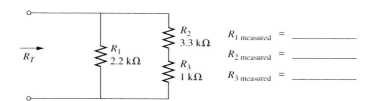

FIG. 9.3

(b) Calculate the total resistance R_T using measured resistor values and insert in Table 9.4.

Calculation:

(c) Use the ohmmeter section of your multimeter to measure the total resistance R_T and record in Table 9.4.

Calculate the magnitude of the percent difference between the calculated value of part 2(b) and the measured value of part 2(c) and insert in Table 9.4.

Calculation:

TABLE 9.4

R_T (calculated)	R_T (measured)	% Difference

(d) If 12 V were applied to the network, as shown in Fig. 9.4, *calculate* the currents I_s, I_1, I_2, and I_3 using measured resistor values and insert in Table 9.5.

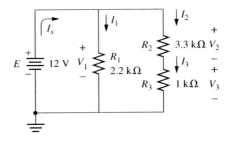

FIG. 9.4

Calculation:

(e) Apply 12 V and measure the currents I_s, I_1, I_2, and I_3 and record in Table 9.5.

Calculate the magnitude of the percent difference between the calculated and measured values for each current and record in Table 9.5.

Calculation:

How are the currents I_2 and I_3 related? Why?

(f) Referring to Fig. 9.4 and Table 9.5 does $I_s = I_1 + I_2$, as required by Kirchhoff's current law?

TABLE 9.5

	Calculated	Measured	%Difference
I_s			
I_1			
I_2			
I_3			

(g) Using the results of part 2(d) and measured resistor values, calculate the voltages V_1, V_2, and V_3 and record in Table 9.6.

Calculation:

(h) Measure the voltages V_1, V_2, and V_3 and record in Table 9.6.

Calculate the magnitude of the percent difference between the calculated and measured values for each voltage and insert in Table 9.6.

Calculation:

TABLE 9.6

	Calculated	Measured	% Difference
V_1			
V_2			
V_3			

(i) How are the voltages E, V_1, and the sum of V_2 and V_3 related? Use Table 9.6 to determine the sum of V_2 and V_3.

Part 3

(a) Construct the series-parallel network of Fig. 9.5 and insert the measured value of each resistor.

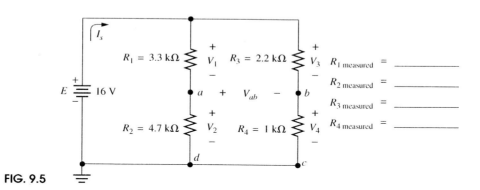

FIG. 9.5

(b) How is the total voltage across the two series elements R_1 and R_2 related to the applied voltage E? Why?

How is the total voltage across the two series elements R_3 and R_4 related to the applied voltage E? Why?

(c) Using the conclusions of part 3(b), calculate the voltages V_2 and V_4 using the voltage divider rule and measured resistor values. Insert the results in Table 9.7.

Calculation:

(d) Measure the voltages V_2 and V_4 and record in Table 9.7.

Calculate the magnitude of the percent difference between calculated and measured values and insert in Table 9.7.

Calculation:

TABLE 9.7

	Calculated	Measured	% Difference
V_2			
V_4			
V_{ab}			
I_s			

(e) Using the results of part 3(c), calculate the voltage V_{ab} using Kirchhoff's voltage law and record in Table 9.7.

Calculation:

(f) Measure the voltage V_{ab} and determine the magnitude of the percent difference between the calculated and measured values and record each in Table 9.7.

Calculation:

(g) Is the voltage V_{ab} also equal to $V_3 - V_1$? Why?

(h) Calculate the current I_s using any method you prefer. Use measured resistor values and record in Table 9.7.

Calculation:

(i) Measure the current I_s and calculate the magnitude of the percent difference between calculated and measured values and record both in Table 9.7.

Calculation:

Part 4

(a) Construct the network of Fig. 9.6. Insert the measured value of each resistor.

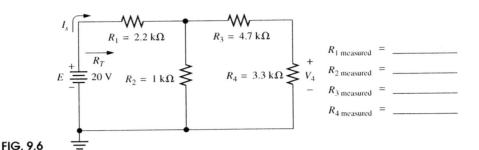

$R_{1\ measured}$ = _____

$R_{2\ measured}$ = _____

$R_{3\ measured}$ = _____

$R_{4\ measured}$ = _____

FIG. 9.6

(b) Calculate the voltage V_4 using the measured resistor values and insert the results in Table 9.8.

Calculation:

(c) Measure the voltage V_4 and calculate the magnitude of the percent difference between calculated and measured values and record in Table 9.8.

Calculation:

(d) Measure the current I_s and calculate the total input resistance from $R_T = E/I_s$ and record both in Table 9.8.

TABLE 9.8

	Calculated	Measured	%Difference
V_4			
I_s			
R_T			

(e) Disconnect the power supply and measure R_T using the ohmmeter section of the DMM. Then calculate the magnitude of the percent difference between the calculated and measured values. Record both results in Table 9.8.

Calculation:

EXERCISES

1. For the series-parallel network of Fig. 9.7, determine V_1, R_1, and R_2 using the information provided. Show all work! Assume $R_{internal} = 0\ \Omega$ for all meters.

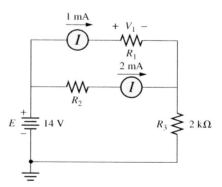

FIG. 9.7

$V_1 =$ _____, $R_1 =$ _____, $R_2 =$ _____

2. For the series-parallel network of Fig. 9.8, determine V_1, R_2, and R_3 using the information provided. Show all work! Assume $R_{internal} = 0\ \Omega$ for all meters.

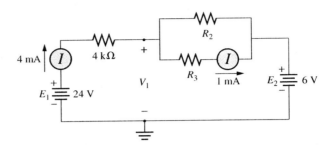

FIG. 9.8

$V_1 =$ _____ , $R_2 =$ _____ , $R_3 =$ _____

3. For the series-parallel network of Fig. 9.5 record the measured values of V_2, V_4, and V_{ab} in Table 9.9 (from Table 9.7).

   Then, using PSpice or Multisim, determine the same voltages and enter the results in Table 9.9. Use measured values for the resistors.

TABLE 9.9

	Table 9.7	PSpice or Multisim
V_2		
V_4		
V_{ab}		

How do the results compare? Try to explain any differences. Attach all appropriate printouts.

Superposition
Theorem (dc)

OBJECTIVES

1. **Validate the superposition theorem.**
2. **Demonstrate that the superposition theorem can be applied to both current and voltage levels.**
3. **Demonstrate that the superposition theorem cannot be applied to nonlinear functions.**

EQUIPMENT REQUIRED

Resistors

1—1.2-kΩ, 2.2-kΩ, 3.3-kΩ, 4.7-kΩ, 6.8-kΩ (1/4-W)

Instruments

1—DMM

2*—dc power supplies

*The unavailability of two power supplies may require that two groups work together.

EQUIPMENT ISSUED

TABLE 10.0

Item	Manufacturer and Model No.	Laboratory Serial No.
DMM		
Power supply		
Power supply		

RÉSUMÉ OF THEORY

The superposition theorem states that the current through, or voltage across, any resistive branch of a multisource network is the algebraic sum of the contributions due to each source acting independently. When the effects of one source are considered, the others are replaced by their internal resistances. Superposition is effective only for linear circuit relationships.

This theorem permits one to analyze circuits without resorting to simultaneous equations. Nonlinear effects, such as power, which varies as the square of the current or voltage, cannot be analyzed using this principle.

PROCEDURE

Part 1 Superposition Theorem (Applied to Current Levels)

The first configuration to be analyzed using the superposition theorem appears in Fig. 10.1. The currents I_1, I_2, and I_3 will be determined by considering the effects of E_1 and E_2 and then adding the resulting levels algebraically.

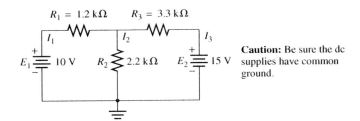

FIG. 10.1

(a) Determining the effects of E_1:

Construct the network of Fig. 10.2 and insert the measured value of each resistor. Note that the supply E_2 has been replaced by a short-circuit equivalent. This *does not mean* that one should place a short-circuit across the terminals of the supply. Simply remove the supply from the network and replace it with a direction to ground, as shown in Fig. 10.2. Keep this in mind for all similar operations throughout the laboratory session. Calculate the currents I_1, I_2, and I_3 using the measured resistor values and insert in Table 10.1.

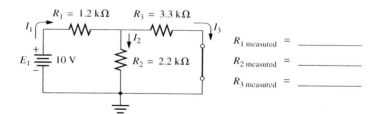

FIG. 10.2

$R_{1 \text{ measured}}$ = _____

$R_{2 \text{ measured}}$ = _____

$R_{3 \text{ measured}}$ = _____

Calculation:

Turn on the supply E_1 and measure the currents I_1, I_2, and I_3. Check your measurements by noting whether Kirchhoff's current law ($I_1 = I_2 + I_3$) is satisfied. Record the results in Table 10.1.

How do the calculated and measured values of I_1, I_2, and I_3 compare?

(b) Determining the effects of E_2:

Construct the network of Fig. 10.3 and insert the measured value of each resistor. Calculate the currents I_1, I_2, and I_3 using the measured resistor values and record in Table 10.1.

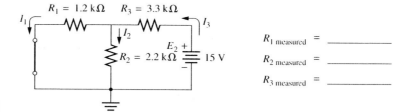

FIG. 10.3

$R_{1 \text{ measured}}$ = _____

$R_{2 \text{ measured}}$ = _____

$R_{3 \text{ measured}}$ = _____

Calculation:

Turn on the supply E_2 and measure the currents I_1, I_2, and I_3. Check your measurements by noting whether Kirchhoff's current law ($I_3 = I_1 + I_2$) is satisfied. Record the results in Table 10.1.

How do the calculated and measured values of I_1, I_2, and I_3 compare?

(c) Determining the total effects of E_1 and E_2:

Construct the network of Fig. 10.1 and insert the measured value of each resistor. Using the calculated results of parts 1(a) and 1(b), calculate the net currents I_1, I_2, and I_3, being very aware of their directions in Figs. 10.2 and 10.3. Next to each result, indicate the direction of the resulting current through each resistor. Record the results in Table 10.1.

TABLE 10.1

	E_1 (part(a))		E_2 (part(b))		E_T (part(c))	
	Calculated	Measured	Calculated	Measured	Calculated	Measured
I_1						
I_2						
I_3						

Turn on both supplies and measure the currents I_1, I_2, and I_3. Determine the direction of each current from the meter connections and insert next to the measured value. Record the measured results in Table 10.1.

How do the calculated and measured levels compare? Has the superposition theorem been validated?

(d) Power levels:

Using the measured current levels of part 1(a) (second column of Table 10.1), calculate the power delivered to each resistor. Show all work! Record the results in Table 10.2.

Calculation:

Using the measured current levels of part 1(b) (fourth column of Table 10.1), calculate the power delivered to each resistor and record in Table 10.2.

Calculation:

Using the measured current levels of part 1(c) (sixth and final column of Table 10.1), calculate the power delivered to each resistor and record in Table 10.2.

Calculation:

TABLE 10.2

	E_1	E_2	E_T	$P_T\,(E_1 + E_2)$
P_1				
P_2				
P_3				

In the final column of Table 10.2, record the power resulting from a simple addition of the powers from the first two columns of the table.

Based on these results, is the superposition theorem applicable to power effects? Explain your answer.

Part 2   Superposition Theorem (Applied to Voltage Levels)

The second configuration to be analyzed using the superposition theorem appears in Fig. 10.4. The voltages V_1, V_2, and V_3 will be determined by considering the effects of E_1 and E_2 and then adding the resulting levels algebraically.

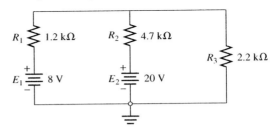

FIG. 10.4

(a) **Determining the effects of E_1:**

Calculate the voltages V_1, V_2, and V_3 for the network of Fig. 10.5 using measured resistor values and insert in Table 10.3. Insert the measured resistor values in the space provided.

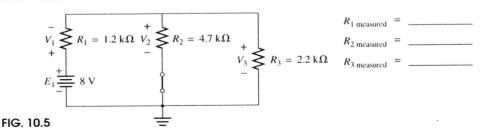

$R_{1\ measured} = \underline{\hspace{3cm}}$

$R_{2\ measured} = \underline{\hspace{3cm}}$

$R_{3\ measured} = \underline{\hspace{3cm}}$

FIG. 10.5

Calculation:

TABLE 10.3

	E_1 (part(a))		E_2 (part(b))		E_T (part(c))	
	Calculated	Measured	Calculated	Measured	Calculated	Measured
V_1						
V_2						
V_3						

Construct the network of Fig. 10.5, turn on the supply E_1, measure the voltages V_1, V_2, and V_3, and record in Table 10.3.

How do the calculated and measured values of V_1, V_2, and V_3 compare?

(b) Determining the effects of E_2:
Calculate the voltages V_1, V_2, and V_3 for the network of Fig. 10.6 using the measured resistor values and record in Table 10.3.

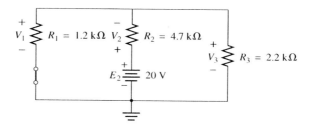

FIG. 10.6

Calculation:

Construct the network of Fig. 10.6, turn on the supply, measure the voltages V_1, V_2, and V_3, and record in Table 10.3.

How do the calculated and measured values of V_1, V_2, and V_3 compare?

(c) Determining the total effects of E_1 and E_2:
Using the calculated results of parts 2(a) and 2(b), calculate the net voltages V_1, V_2, and V_3 and record in Table 10.3. Be very aware of their polarities in Figs. 10.5 and 10.6. Indicate the polarity of the voltage across each resistor on Fig. 10.4.

Calculation:

Construct the network of Fig. 10.4, turn on the supply, measure the voltages V_1, V_2, and V_3, and record in Table 10.3. Be sure to note the polarity of each reading on the schematic.

How do the calculated and measured levels compare? Has the superposition theorem been validated for voltage levels?

Part 3   A Third Configuration

(For this part you must have supplies with isolated ground connections. If not available, do not complete this part.)

(a) Construct the network of Fig. 10.7, taking special note of the fact that the positive side of E_2 is connected to ground potential. Using the measured resistor values, calculate the voltages V_1, V_2, and V_3 for each source using superposition and insert in Table 10.4. Then calculate the total effect of the two sources and insert in Table 10.4. Show your work in the space provided. Indicate the resulting polarities for each voltage next to each result.

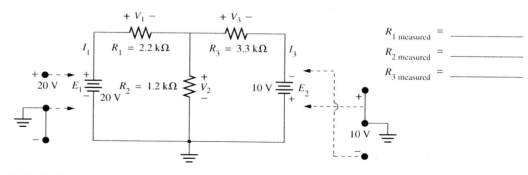

FIG. 10.7

Calculation:

TABLE 10.4

	E_1 (calculated)	E_2 (calculated)	$E_1 + E_2$ (calculated)	Measured
V_1				
V_2				
V_3				

(b) Construct and energize the network of Fig. 10.7, measure the voltages V_1, V_2, and V_3, and record in last column of Table 10.4. Is the superposition theorem verified?

EXERCISES

1. For the network of Fig. 10.8:

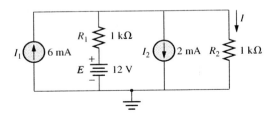

FIG. 10.8

(a) By inspection (meaning no calculations whatsoever) using the superposition theorem, which source (I_1, I_2, or E) would appear to have the most impact on the current, I?

(b) Determine the current, I, using superposition and note whether your conclusion in part (a) was correct.

$I =$ _____

2. Using superposition, determine the current, I, for the network of Fig. 10.9.

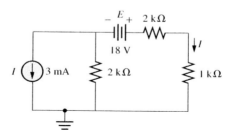

FIG. 10.9

$I = $ _____

3. For the network of Fig. 10.1, record the measured values of I_1, I_2, and I_3 for each source and for both sources from Table 10.1 into Table 10.5. Then, using PSpice or Multisim, determine the currents for each source and for both sources and record the results in Table 10.5.

TABLE 10.5

	E_1 (Table 10.1)	E_1 (PSpice or Multisim)	E_2 (Table 10.2)	E_2 (PSpice or Multisim)	E_T (Table 10.1)	E_T (PSpice or Multisim)
I_1						
I_2						
I_3						

Compare results for each source and for both sources. Try to explain any major differences. Attach all appropriate printouts.

EXPERIMENT dc

11

Thevenin's Theorem and Maximum Power Transfer

OBJECTIVES

1. Validate Thevenin's theorem through experimental measurements.
2. Become aware of an experimental procedure to determine E_{Th} and R_{Th}.
3. Demonstrate that maximum power transfer to a load is defined by the condition $R_L = R_{Th}$.

EQUIPMENT REQUIRED

Resistors

1—91-Ω, 220-Ω, 330-Ω, 470-Ω, 1-kΩ, 2.2-kΩ, 3.3-kΩ (1/4-W)

1—0–1-kΩ potentiometer, 0–10-kΩ potentiometer

Instruments

1—DMM

1—dc power supply

EQUIPMENT ISSUED

TABLE 11.0

Item	Manufacturer and Model No.	Laboratory Serial No.
DMM		
Power supply		

RÉSUMÉ OF THEORY

Through the use of Thevenin's theorem, a complex two-terminal, linear, multisource dc network can be replaced by one having a single source and resistor.

The Thevenin equivalent circuit consists of a single dc source referred to as the *Thevenin voltage* and a single fixed resistor called the *Thevenin resistance*. The Thevenin voltage is the open-circuit voltage across the terminals in question (Fig. 11.1). The Thevenin resistance is the resistance between these terminals with all of the voltage and current sources replaced by their internal resistances (Fig. 11.1).

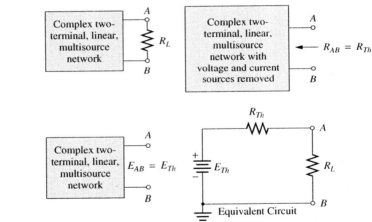

FIG. 11.1

If a dc voltage source is to deliver maximum power to a resistor, the resistor must have a value equal to the internal resistance of the source. In a complex network, maximum power transfer to a load will occur when the load resistance is equal to the Thevenin resistance "seen" by the load. For this value, the voltage across the load will be one-half of the Thevenin voltage. In equation form,

$$R_L = R_{Th}, \quad V_L = \frac{E_{Th}}{2}, \quad \text{and} \quad P_{max} = \frac{E_{Th}^2}{4R_{Th}} \tag{11.1}$$

PROCEDURE

Part 1   Thevenin's Theorem

Calculation:

(a) Construct the network of Fig. 11.2. Calculate the Thevenin voltage and resistance for the network to the left of points a–a' using measured resistor values. Enter the results in Table 11.1. Show all work!

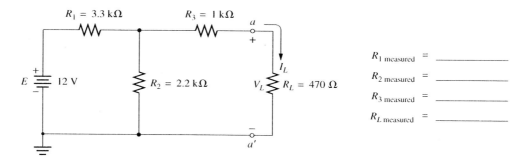

$R_{1\ measured} =$ _____

$R_{2\ measured} =$ _____

$R_{3\ measured} =$ _____

$R_{L\ measured} =$ _____

FIG. 11.2

Calculation:

TABLE 11.1

	Calculated Values [Part 1(a)]	Measured Values [Parts 1(e) and 1(f)]	% Difference
E_{Th}			
R_{Th}			

(b) Insert the values of E_{Th} and R_{Th} in Fig. 11.3 and calculate I_L. Record the results in Table 11.2.

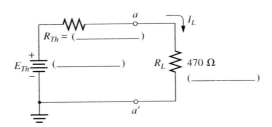

FIG. 11.3

Calculation:

(c) Calculate the current I_L in the original network of Fig. 11.2 using series-parallel techniques (use measured resistor values) and insert in Table 11.2. Show all work!

Calculation:

TABLE 11.2

I_L (equivalent)	I_L (series-parallel)

How does this calculated value of I_L compare to the value of part 1(b)?

Measurements:

      (d) Turn on the 12-V supply of Fig. 11.2, measure the voltage V_L, and record in Table 11.3. Using the measured value of R_L, calculate the current I_L and record in Table 11.3.

Calculation:

      How does this measured value of I_L compare with the calculated levels of parts 1(b) and 1(c)?

Determining R_{Th}:

      (e) Determine R_{Th} by constructing the network of Fig. 11.4 and measuring the resistance between points a–a' with R_L removed. Record in Table 11.1.

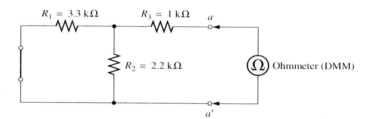

FIG. 11.4

Determining E_{Th}:

      (f) Determine E_{Th} by constructing the network of Fig. 11.5 and measuring the open-circuit voltage between points a–a'. Record the result in Table 11.1.

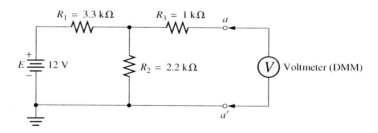

FIG. 11.5

Thevenin Network:

(g) Construct the network of Fig. 11.6 and set the values obtained for the measured values of E_{Th} and R_{Th} in parts 1(e) and 1(f), respectively. Use the ohmmeter section of your meter to set the potentiometer properly. Then measure the voltage V_L and calculate the current I_L using the measured value of R_L. Record both results in Table 11.3.

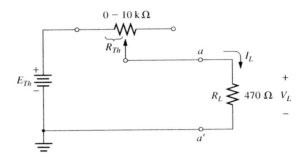

FIG. 11.6

TABLE 11.3

	V_L	I_L(from V_L)
Original network		
Thevenin equivalent		

How does this value of I_L compare with the calculated level of part 1(b)?

How do the calculated and measured values of E_{Th} and R_{Th} compare? Insert the magnitude of the percent difference in the third column of Table 11.1 using the equations:

$$\% \text{ Difference} = \left| \frac{\text{Measured} - \text{Calculated}}{\text{Measured}} \right| \times 100\% \qquad (11.2)$$

Calculation:

Noting the overall results of Table 11.1, has Thevenin's theorem been verified?

Part 2 Maximum Power Transfer (Validating the Condition $R_L = R_{Th}$)

(a) Construct the network of Fig. 11.7 and set the potentiometer to 50 Ω. Measure the voltage across R_L as you vary R_L through the following values: 50, 100, 200, 300, 330, 400, 600, 800, and 1000 Ω. Be sure to set the resistance with the ohmmeter section of your meter before each reading. When setting the resistance level, remember to turn off the dc supply and disconnect one terminal of the potentiometer. Complete Table 11.3 and plot P_L versus R_L on Graph 11.1.

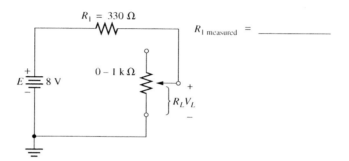

FIG. 11.7

TABLE 11.4

R_L	V_L	$P = \dfrac{V_L^2}{R_L}$ (mW)
0 Ω	0 V	0 mW
50 Ω		
100 Ω		
200 Ω		
300 Ω		
$R_{1\,measured} =$ _____		
400 Ω		
600 Ω		
800 Ω		
1000 Ω		

(b) Theoretically, for the network of Fig. 11.7, what value of R_L should result in maximum power to R_L? Record in Table 11.5.

TABLE 11.5

	Theory	Experimental
R_L		
V_L		

Referring to the plot of Graph 11.1, what value of R_L resulted in maximum power transfer to R_L? Record in Table 11.5.

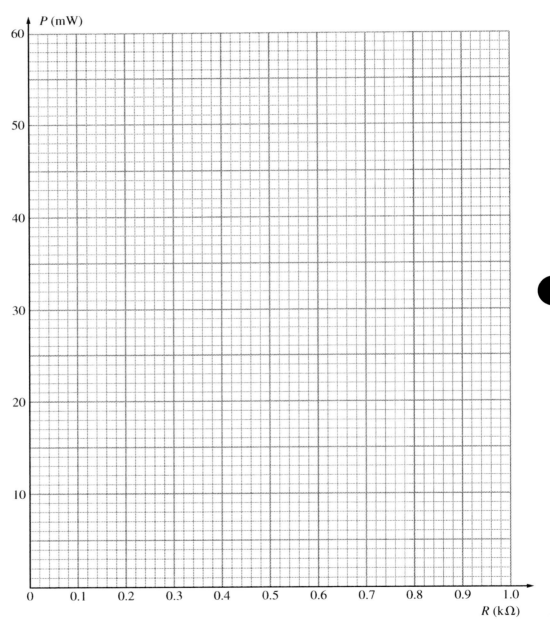

GRAPH 11.1

How do the theoretical and measured values of R_L compare?

(c) Under maximum power transfer conditions, how are the voltages V_L and E related? Why?

Based on the preceding conclusion, determine V_L for maximum power transfer to R_L and record in Table 11.5.

Set the potentiometer to the resistance R_L that resulted in maximum power transfer on Graph 11.1 and measure the resulting voltage across R_L. Record in Table 11.5.

How does the measured value compare to the expected theoretical level?

Part 3   Maximum Power Transfer (Experimental Approach)

(a) Construct the network of Fig. 11.8. Insert the measured value of each resistor.

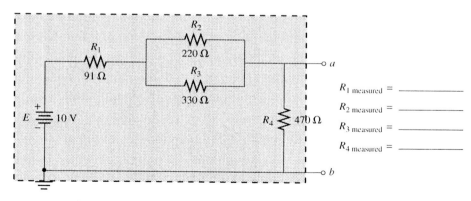

$R_{1\ measured}$ = _____

$R_{2\ measured}$ = _____

$R_{3\ measured}$ = _____

$R_{4\ measured}$ = _____

FIG. 11.8

(b) The Thevenin equivalent circuit will now be determined for the network to the left of the terminals a–b without disturbing the structure of the network. All the measurements will be made at the terminals a–b.

E_{Th}:

Determine E_{Th} by turning on the supply and measuring the open-circuit voltage V_{ab}. Record the reading in Table 11.6.

R_{Th}:

Introduce the 1-kΩ potentiometer to the network of Fig. 11.8, as shown in Fig. 11.9.

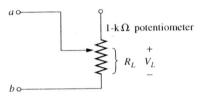

FIG. 11.9

Turn on the supply and adjust the potentiometer until the voltage V_L is $E_{Th}/2$, a condition that must exist if $R_L = R_{Th}$. Then turn off the supply and remove the potentiometer from the network without disturbing the position of the wiper arm. Measure the resistance between the two terminals connected to a–b and record as R_{Th} in Table 11.6, first column.

TABLE 11.6

	Measured	Calculated
E_{Th}		
R_{Th}		

(c) Now we need to check our measured results against a theoretical solution. Calculate R_{Th} and E_{Th} for the network to the left of terminals a–b of Fig. 11.8 and record in the second column of Table 11.6. Use measured resistor values.

Calculation:

How do the calculated and measured values compare?

(d) Now plot P_L and V_L versus R_L to confirm once more that the conditions for maximum power transfer to a load are that $R_L = R_{Th}$ and $V_L = E_{Th}/2$.

Leave the potentiometer as connected in Fig. 11.9 and measure V_L for all the values of R_L appearing in Table 11.7. Then calculate the resulting power to the load and complete the table. Finally, plot both P_L and V_L versus R_L on Graphs 11.2 and 11.3, respectively.

TABLE 11.7

R_L^*	V_L(measured)	$P_L = V_L^2/R_L$ (calculated)
0 Ω	0 V	0 mW
25 Ω		
50 Ω		
100 Ω		
150 Ω		
200 Ω		
250 Ω		
300 Ω		
350 Ω		
400 Ω		
450 Ω		
500 Ω		

*Be sure to remove the potentiometer from the network when setting each value of R_L.
At the very least, disconnect one side of the potentiometer when making the setting.

Reviewing Graph 11.2, did maximum power transfer to the load occur when $R_L = R_{Th}$? What conclusion can be drawn from the results?

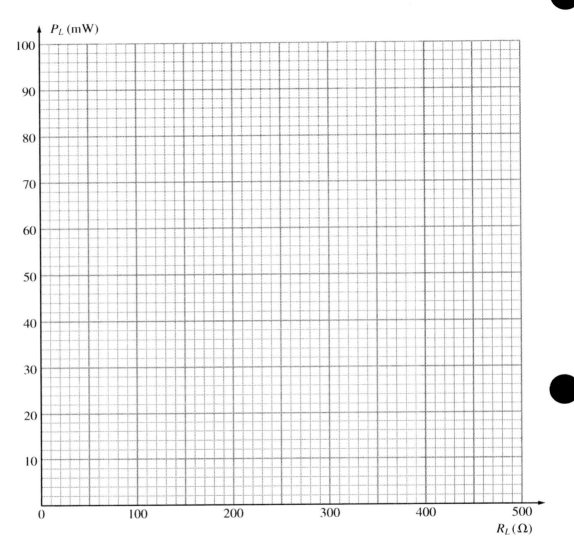

GRAPH 11.2

Noting Graph 11.3, does $V_L = E_{Th}/2$ when $R_L = R_{Th}$? Comment accordingly.

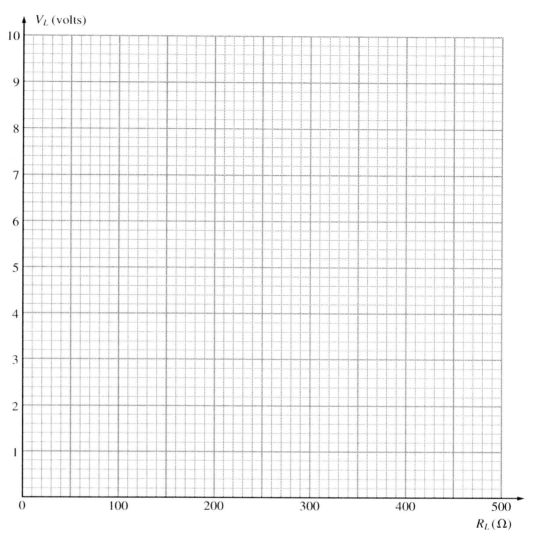

GRAPH 11.3

EXERCISES

1. For the network of Fig. 11.10:

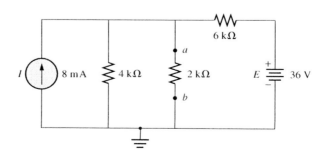

FIG. 11.10

(a) Determine R_{Th} and E_{Th} for the network external to the 2-kΩ resistor.

$$R_{Th} = \underline{\hspace{2cm}}, \; E_{Th} = \underline{\hspace{2cm}}$$

(b) Determine the power delivered to the 2-kΩ resistor using the Thevenin equivalent circuit.

(c) Is the power determined in part (b) the maximum power that could be delivered to a resistor between terminals a and b? If not, what is the maximum power?

2. For the network of Fig. 11.2 record the measured values of E_{Th} and R_{Th} from Table 11.1 in Table 11.8. Then, using PSpice or Multisim, and methods described in the text *Introductory Circuit Analysis*, determine each quantity and record in Table 11.8.

TABLE 11.8

	Table 11.1	PSpice or Multisim
E_{Th}		
R_{Th}		

Compare results for each quantity and try to explain any major differences. Attach all appropriate printouts.

Norton's Theorem and Current Sources

OBJECTIVES

1. Validate Norton's theorem through experimental measurements.
2. Become aware of an experimental procedure to determine I_N and R_N.
3. Demonstrate how a current source can be constructed (for a limited load range) using a voltage source.

EQUIPMENT REQUIRED

Resistors

1—10-Ω, 47-Ω, 100-Ω, 220-Ω, 330-Ω, 3.3-kΩ, 10-kΩ (1/4-W)

1—0–1-kΩ potentiometer

Instruments

1—DMM

1—dc power supply

EQUIPMENT ISSUED

TABLE 12.0

Item	Manufacturer and Model No.	Laboratory Serial No.
DMM		
Power supply		

RÉSUMÉ OF THEORY

Through the use of Thevenin's or Norton's theorem, a complex two-terminal, linear, multisource dc network can be replaced by a single source and resistor.

The Norton equivalent circuit is a single dc current source in parallel with a resistor. The *Norton current* (I_N) is the short-circuit current between the terminals in question. The *Norton resistance* (R_N) is the resistance between these terminals with all voltage and current sources replaced by their internal resistances.

The theory of source conversion dictates that the Norton and Thevenin circuits be terminally equivalent and related as follows:

$$\boxed{R_N = R_{Th}} \qquad \boxed{E_{Th} = I_N R_N} \qquad \boxed{I_N = \frac{E_{Th}}{R_{Th}}} \tag{12.1}$$

A current source of some practical value can be constructed using a voltage source if the source resistance (R_s), as shown in Fig. 12.1(a), is much greater than the range of the load resistance R_L.

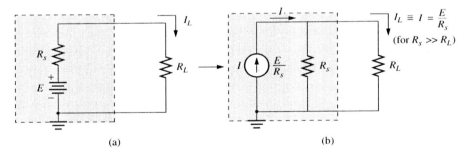

FIG. 12.1

For Fig. 12.1(a)

$$I_L = \frac{E}{R_s + R_L}$$

If $R_s \gg R_L$, $R_s + R_L \cong R_s$ and

$$I_L \cong \frac{E}{R_s}$$

demonstrating that the current I_L of the load is essentially unaffected by changes in R_L as long as $R_s \gg R_L$. The equivalent current source for $R_s \gg R_L$ appears in Fig. 12.1(b).

PROCEDURE

Part 1 Designing a 1-mA Current Source

(**a**) Construct the network of Fig. 12.2. Insert the measured value of R_s in Fig. 12.2.

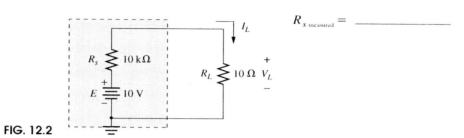

$R_{s\text{ measured}} = $ _____

FIG. 12.2

(**b**) Record the measured R_L value in the first column of Table 12.1. Using measured resistor values, calculate the current I_L and record in Table 12.1.

Calculation:

(**c**) Turn on the supply and measure the voltage V_L. Then calculate the current I_L using the measured resistor value. Record the values in Table 12.1.

Calculation:

(**d**) Turn off the supply and change R_L to 47 Ω. Record the measured resistor value and calculate the resulting current I_L. Record the new value of I_L in Table 12.1.

Calculation:

(e) Turn on the supply and measure the voltage V_L. Then calculate the current I_L using the measured resistor value. Record both results in Table 12.1.

TABLE 12.1

	10 Ω	47 Ω	100 Ω	220 Ω
R_L (measured)				
I_L (calculated)				
V_L (measured)				
I_L (from V_L)				

Calculation:

(f) Increase R_L to 100 Ω. Record the measured resistor value and calculate the resulting current I_L. Record the new value of I_L in Table 12.1.

Calculation:

(g) Turn on the supply and measure the voltage V_L. Then calculate the current I_L using the measured resistor value. Record both results in Table 12.1.

Calculation:

(h) Increase R_L to 220 Ω. Record the measured resistor value and calculate the resulting current I_L. Record the new value of I_L in Table 12.1.

Calculation:

(i) Turn on the supply and measure the voltage V_L. Then calculate the current I_L using the measured resistor value. Record both results in Table 12.1.

Calculation:

(j) Plot I_L versus measured R_L on Graph 12.1 for the values of I_L calculated from the measured voltage V_L.

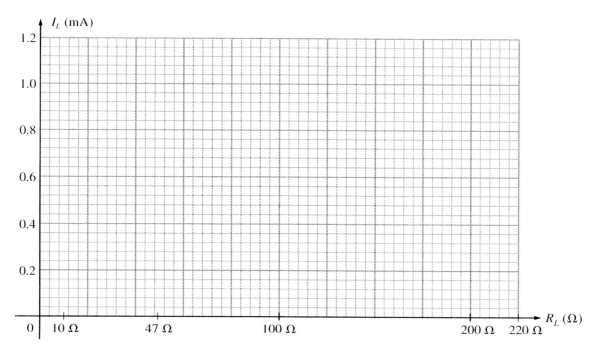

GRAPH 12.1

(k) Does the plot of Graph 12.1 reveal I_L is fairly constant for the range of chosen R_L values?

(l) Would you consider the network of Fig. 12.3 a good approximation to the configuration of Fig. 12.2? Make any appropriate comments.

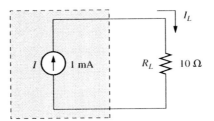

FIG. 12.3

Conclusion:

Part 2   Determining R_N and I_N

(a) Construct the network of Fig. 12.4(a). Insert the measured value of each resistor.

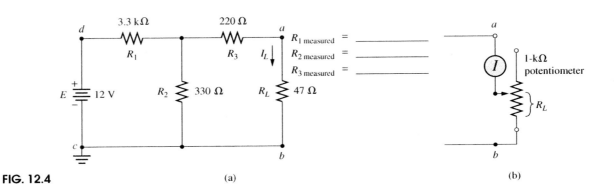

FIG. 12.4   (a)   (b)

(b) Using measured resistor values, calculate the levels of R_N and I_N for the network to the left of the 47-Ω resistor and record the results in the first column of Table 12.2.

(c) Using the Norton equivalent circuit calculated in part 2(b), calculate the current I_L for a load of 47 Ω.

Calculation:

$I_L = $ _____

TABLE 12.2

	Calculated (b)	Measured (e) & (f)
R_N		
I_N		

(d) Turn on the supply of Fig. 12.4 and measure the voltage V_{ab}. Then calculate the current I_L using the measured value of the 47 Ω resistor. Record both values in the first column of Table 12.3.

Calculation:

TABLE 12.3

	47 Ω (Fig. 12.4)	47 Ω (Norton Equiv.)	100 Ω (Norton Equiv.)	100 Ω (Fig. 12.4)
V_{ab}				
I_L				

How do the calculated (part 2(c)) and measured values of I_L compare?

(e) The level of I_N can be determined by replacing the 47-Ω resistor by a short circuit and measuring the short-circuit current. Since the internal resistance of an ammeter is relatively low, all this can be accomplished by simply removing the 47-Ω resistor and replacing it with the ammeter section of the DMM. Since the current level is unknown, start with the highest scale and work down to the scale that provides the highest degree of accuracy. Record the level of I_N in the second column of Table 12.2.

How does the calculated value of I_N from part 2(b) compare with this measured value of I_N?

(f) R_N is now determined experimentally by first calculating $I_N/2$ using the measured value from part 2(e). For the Norton equivalent circuit with $R_L = R_N$, $I_L = I_N/2$.

$$\frac{I_N}{2} = \underline{\hspace{3cm}}$$

Connect the 1-kΩ potentiometer and ammeter in a series configuration between points a and b, as shown in Fig. 12.4(b). Turn on the supply and vary the potentiometer until the ammeter reading is $I_N/2$. Then remove the potentiometer and measure R_N and record in Table 12.2.

How does the calculated value of R_N from part 2(b) compare with this measured value of R_N?

(g) We will now construct the Norton equivalent circuit defined by the calculated levels of R_N and I_N from part 2(b).

First construct the network of Fig. 12.5.

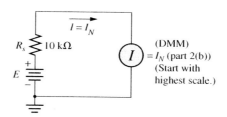

FIG. 12.5

Then vary the supply voltage until the DMM indicates the value I_N from part 2(b). Record the values of E and I_N in Fig. 12.6(b). Next remove the DMM and, using it as an ohmmeter, set the 0–1-kΩ potentiometer to the value of R_N from part 2(b). Now insert the 0–1-kΩ potentiometer in the circuit of Fig. 12.6(b).

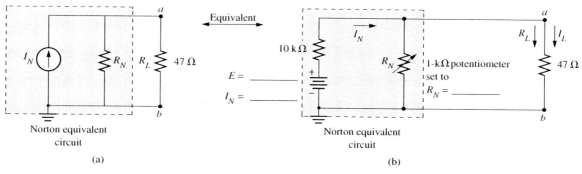

FIG. 12.6

The network of Fig. 12.6(b) is the Norton equivalent circuit. The 0–1-kΩ potentiometer is equivalent to R_N, and the 10-kΩ resistor in series with the power supply is the equivalent current source. The 10-kΩ resistor was chosen to ensure minimum sensitivity on the part of I_N to the smaller resistor values connected in parallel in Fig. 12.6(b). In other words, $I_N = E/(10 \text{ k}\Omega + R_{network}) \cong E/10 \text{ k}\Omega$ and, therefore, approximates an ideal current source.

Measure the voltage V_{ab} and compute I_L using the measured resistor value and record in the second column of Table 12.3.

Calculation:

How does the level of I_L determined here compare with the calculated level of part 2(c)? Has the Norton equivalent circuit been verified?

(h) Replace the 47-Ω resistor of Fig. 12.6(b) by a 100-Ω resistor and measure the voltage V_{ab}. Then calculate the resulting current I_L using the measured resistor value. Record the results in the third column of Table 12.3.

Calculation:

Reconstruct the network of Fig. 12.4 with $R_L = 100\ \Omega$ and measure the voltage V_{ab}. Then calculate the current I_L using the measured resistor value and record all results in the last column of Table 12.3.

Calculation:

How do the levels of I_L for this part compare? Have we verified that the Norton equivalent circuit is valid for changing loads across its terminals?

Part 3   Source Conversion

(a) Using source conversion techniques, the Norton equivalent circuit of Fig. 12.6(b) can be converted to a Thevenin equivalent circuit. Further, if the circuits are completely equivalent, the measured values of V_{ab} and I_L should be the same as before. The conversion can be accomplished by first calculating the Thevenin voltage E_{Th}. Determine E_{Th} using the conversion equation $E_{Th} = I_N R_N$ and the parameters of Fig. 12.6(b).

$E_{Th} = $ _____

Since $R_{Th} = R_N$, the Thevenin equivalent circuit can now be constructed.

(b) Construct the Thevenin equivalent circuit of Fig. 12.7:

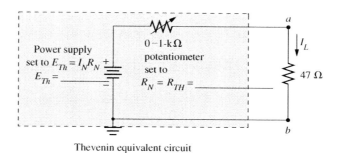

FIG. 12.7   Thevenin equivalent circuit

Now measure the voltage V_{ab} and again compute I_L. (Use the measured value of the 47-Ω resistor.) Record in Table 12.4.

Calculation:

How do V_{ab} and I_L compare to those in part 2(d)? Calculate the percent difference from

$$\% \text{ Difference} = \frac{|\text{Part 2(d)} - \text{Part 3(b)}|}{\text{Part 2(d)}} \times 100\%$$

and record in Table 12.4.

Calculation:

TABLE 12.4

	Measured	% Difference
V_{ab}		
I_L		

Do the results verify the source conversion theory?

EXERCISES

For the network of Fig. 12.8:

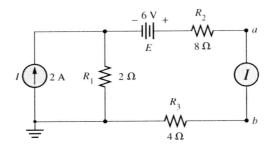

FIG. 12.8

(a) Find I_N by determining the reading of an ammeter placed between points a and b.

$I_N = $ _____

(b) Find R_N by determining the load resistor R_L that must be placed between points a and b to establish $I_L = I_N/2$.

$R_N = $ _____

Experiment dc **13**

Methods of Analysis

OBJECTIVES

1. Validate the branch-current-analysis technique through experimental measurements.
2. Test the mesh- (loop-) analysis approach with experimental measurements.
3. Demonstrate the validity of the nodal-analysis technique through experimental measurements.

EQUIPMENT REQUIRED

Resistors

1—1.0-kΩ, 1.2-kΩ (1/4-W)

2—2.2-kΩ, 3.3-kΩ (1/4-W)

Instruments

1—DMM

2*—dc Power supplies

---

*The unavailability of two supplies will simply mean that two groups must work together.

EQUIPMENT ISSUED

TABLE 13.0

Item	Manufacturer and Model No.	Laboratory Serial No.
DMM		
Power supply		
Power supply		

RÉSUMÉ OF THEORY

The branch-, mesh-, and nodal-analysis techniques are used to solve complex networks with a single source or networks with more than one source that are not in series or parallel.

The branch- and mesh-analysis techniques will determine the currents of the network, while the nodal-analysis approach will provide the potential levels of the nodes of the network with respect to some reference.

The application of each technique follows a sequence of steps, each of which will result in a set of equations with the desired unknowns. It is then only a matter of solving these equations for the various variables, whether they be current or voltage.

PROCEDURE

Part 1   Branch-current Analysis

(a) Construct the network of Fig. 13.1 and insert the measured values of the resistors in the spaces provided.

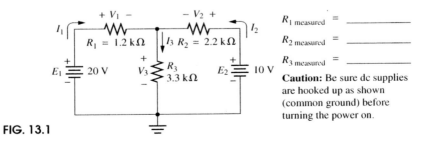

$R_{1\text{ measured}}$ = _____

$R_{2\text{ measured}}$ = _____

$R_{3\text{ measured}}$ = _____

Caution: Be sure dc supplies are hooked up as shown (common ground) before turning the power on.

FIG. 13.1

(b) Using branch-current analysis, *calculate* the current through each branch of the network of Fig. 13.1 and insert in Table 13.1. Use the measured resistor values and assume the current directions shown in the figure. Show all your calculations in the space provided and be neat!

Calculation:

TABLE 13.1

Current	Calculated	Measured	% Difference
I_1			
I_2			
I_3			

(c) Measure the voltages V_1, V_2, and V_3 and enter below with a minus sign for any polarity that is opposite to that in Fig. 13.1. Record the results in Table 13.2.

TABLE 13.2

	Measured Value
V_1	
V_2	
V_3	

Calculate the currents I_1, I_2, and I_3 using the measured resistor values and insert in Table 13.1 as the measured values. Be sure to include a minus sign if the current direction is opposite to that appearing in Fig. 13.1. Show all work.

Calculation:

How do the calculated and measured results compare? Determine the percent difference for each current in Table 13.1 using the equation:

$$\% \text{ Difference} = \left| \frac{\text{Measured} - \text{Calculated}}{\text{Measured}} \right| \times 100\% \qquad (13.1)$$

Calculation:

Part 2   Mesh Analysis

(a) Construct the network of Fig. 13.2 and insert the measured values of the resistors in the spaces provided.

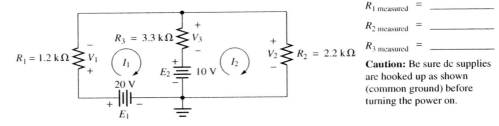

R_1 measured $=$ _____

R_2 measured $=$ _____

R_3 measured $=$ _____

Caution: Be sure dc supplies are hooked up as shown (common ground) before turning the power on.

FIG. 13.2

(b) Using mesh analysis, *calculate* the mesh currents I_1 and I_2 of the network. Use the measured resistor values and the indicated directions for the mesh currents. Then determine the current through each resistor and insert in Table 13.3 in the "Calculated" column. Include all your calculations and organize your work.

Calculation:

TABLE 13.3

Current	Calculated	Measured	% Difference
$I_{R_1} = I_1$			
$I_{R_2} = I_2$			
$I_{R_3} = I_1 - I_2$			

(c) Measure the voltages V_1, V_2, and V_3 and enter in Table 13.4 with a minus sign for any polarity that is opposite to that in Fig. 13.2.

TABLE 13.4

	Measured Value
V_1	
V_2	
V_3	

Calculate the currents I_{R_1}, I_{R_2}, and I_{R_3} using the measured voltage and resistor values and insert in Table 13.3 as the measured values. Be sure to include a minus sign if the current direction is opposite to that defined by the polarity of the voltage across the resistor.

Calculation:

How do the calculated and measured results compare? Determine the percent difference for each current of Table 13.3.

Calculation:

Part 3   Nodal Analysis

(a) Construct the network of Fig. 13.3 and insert the measured resistor values.

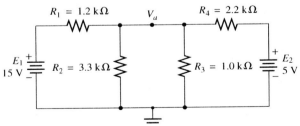

$R_1 = 1.2\ k\Omega$ V_a $R_4 = 2.2\ k\Omega$

E_1 15 V $R_2 = 3.3\ k\Omega$ $R_3 = 1.0\ k\Omega$ E_2 5 V

$R_{1\ measured}$ = _____

$R_{2\ measured}$ = _____

$R_{3\ measured}$ = _____

$R_{4\ measured}$ = _____

Caution: Be sure dc supplies are hooked up as shown (common ground) before turning the power on.

FIG. 13.3

Calculation:

(b) Using measured resistor values, determine V_a using nodal analysis and record in Table 13.5. Show all work and be neat!

TABLE 13.5

	Calculated	Measured
V_a		

(c) Using V_a, calculate the currents I_{R_1} and I_{R_3} using measured resistor values and insert in Table 13.6.

Calculation:

TABLE 13.6

Current	Calculated	Measured	% Difference
I_{R_1}			
I_{R_3}			

Measurements:

(d) Energize the network and measure the voltage V_a. Insert the result in Table 13.5. Compare with the result of part 3(b).

(e) Using V_a(measured), calculate the currents I_{R_1} and I_{R_3} using measured resistor values and insert in Table 13.6 as the measured results.

Calculation:

(f) How do the calculated and measured results for I_{R_1} and I_{R_3} compare? Determine the percent differences for each current and insert in Table 13.6.

Calculation:

Part 4   Bridge Network

(a) Construct the network of Fig. 13.4. Insert the measured resistor values.

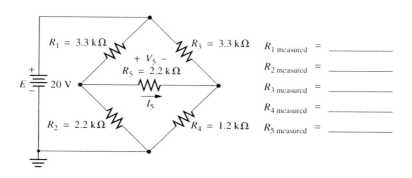

FIG. 13.4

(b) Using any one of the three techniques examined in this experiment, calculate the voltage V_5 and the current I_5. Use the measured resistor values. Record the results in the "Calculated" column of Table 13.7.

Calculation:

TABLE 13.7

	Calculated	Measured	% Difference
V_5			
I_5			

(c) Measure the voltage V_5 and insert in the "Measured" column of Table 13.7 with a minus sign if the polarity is different from that appearing in. Fig. 13.4.

(d) Calculate the percent difference between the two values of V_5. Record the result in Table 13.7.

Calculation:

(e) Calculate the current I_5 using the measured value of V_5 and the measured value of the resistor R_5. Record in Table 13.7.

Calculation:

Determine the percent difference and record in Table 13.7.

Calculation:

EXERCISES

1. Many times one is faced with the question of which method to use in a particular problem. The laboratory activity does not prepare one to make such choices but only shows that the methods work and are solid. From your experience in this activity, summarize in your own words which method you prefer and why you chose the method you did for the analysis of part 4.

2. For the network of Fig. 13.1, record the measured values of I_1, I_2, and I_3 from Table 13.1 to Table 13.8. Then, using PSpice or Multisim, determine the branch currents and record in Table 13.8.

TABLE 13.8

Table 13.1	PSpice or Multisim
I_1	
I_2	
I_3	

Compare the results and comment accordingly. Attach all appropriate printouts.

Capacitors

OBJECTIVES

1. **Validate conclusions regarding the behavior of capacitors in a steady-state dc network.**
2. **Plot the exponential curve for the voltage across a charging capacitor.**
3. **Verify the basic equations for determining the total capacitance for capacitors in series and parallel.**
4. **Demonstrate the usefulness of Thevenin's theorem for networks not having the basic series *R-C* form.**

EQUIPMENT REQUIRED

Resistors

2—1.2-kΩ, 100-kΩ (1/4-W)
1—3.3-kΩ, 47-kΩ (1/4-W)

Capacitors

1—100-μF, 220-μF (electrolytic)

Instruments

1—DMM
1—dc Power supply
1—Single-pole, single-throw switch

EQUIPMENT ISSUED

TABLE 14.0

Item	Manufacturer and Model No.	Laboratory Serial No.
DMM		
Power supply		

RÉSUMÉ OF THEORY

The resistor dissipates electrical energy in the form of heat. In contrast, the capacitor is a component that stores electrical energy in the form of an electrical field. The capacitance of a capacitor is a function of its geometry and the dielectric used. Dielectric materials (insulators) are rated by their ability to support an electric field in terms of a figure called the dielectric constant (k). Dry air is the standard dielectric for purposes of reference and is assigned the value of unity. Mica, with a dielectric constant of five, has five times the capacitance of a similarly constructed air capacitor.

One of the most common capacitors is the parallel-plate type. The capacitance (in farads) is determined by

$$C = 8.85 \times 10^{-12} \epsilon_r \frac{A}{d} \qquad \text{(farads, F)} \qquad (14.1)$$

where ϵ_r is the relative permittivity (or dielectric constant), A is the area of the plates (m^2), and d is the distance between the plates (m). By changing any one of the three parameters, one can easily change the capacitance.

The dielectric constant is not to be confused with the dielectric strength of a material, which is given in volts per unit length and is a measure of the maximum stress that a dielectric can withstand before it breaks down and loses its insulator characteristics.

In a dc circuit, the volt-ampere characteristics of a capacitor in the steady-state mode are such that the capacitor prevents the flow of dc current but will charge up to a dc voltage. Essentially, therefore, the characteristics of a capacitor in the steady-state mode are those of an *open circuit*. The charge Q stored by a capacitor is given by

$$Q = CV \qquad (14.2)$$

where C is the capacitance and V is the voltage impressed across the capacitor.

Capacitors in series are treated like resistors in parallel:

$$\frac{1}{C_T} = \frac{1}{C_1} + \frac{1}{C_2} + \frac{1}{C_3} + \cdots + \frac{1}{C_N} \qquad (14.3)$$

Capacitors in parallel are treated like resistors in series:

$$C_T = C_1 + C_2 + C_3 + \cdots + C_N \qquad (14.4)$$

The energy stored by a capacitor (in joules) is determined by

$$W = \frac{1}{2} CV^2 \qquad (14.5)$$

For the network of Fig. 14.1, the capacitor will, for all practical purposes, charge up to E volts in *five* time constants, where a time constant (τ) is defined by

$$\boxed{\tau = RC}$$ (14.6)

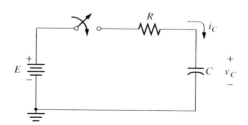

FIG. 14.1

In one time constant, the voltage v_C will charge up to 63.2% of its final value, in 2τ up to 86.5%, in 3τ up to 95.1%, in 4τ up to 98.1%, and in 5τ up to 99.3%, as defined by

$$\boxed{v_C = E(1 - e^{-t/RC})}$$ (14.7)

The current i_C is defined by

$$\boxed{i_C = \frac{E}{R}(e^{-t/RC})}$$ (14.8)

PROCEDURE

Part 1 Basic Series *R-C* Circuit

(a) Construct the network of Fig. 14.2. Insert the measured resistor value. Be sure to note polarity on electrolytic capacitors as shown in the figure.

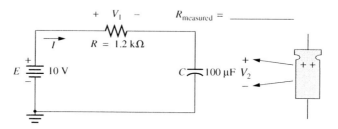

FIG. 14.2

(b) Calculate the steady-state value (defined by a period of time greater than five time constants) of the current I and the voltages V_1 and V_2. Record in Table 14.1.

Calculation:

TABLE 14.1

	Calculated	Measured
I		
V_1		
V_2		

(c) Measure the voltages V_1 and V_2 and calculate the current I from Ohm's law and insert in Table 14.1. Compare with the results of part 1(b).

Calculation:

(d) Calculate the energy stored by the capacitor.

Calculation:

$W =$ _____

(e) Carefully disconnect the supply and quickly measure the voltage across the disconnected capacitor. Is there a reading? Why?

$V_C =$ _____

(f) Short the capacitor terminals with a lead and then measure V_C again. Why was it necessary to perform this step?

Part 2   Parallel *R-C* dc Network

(a) Construct the network of Fig. 14.3. Insert the measured resistor values.

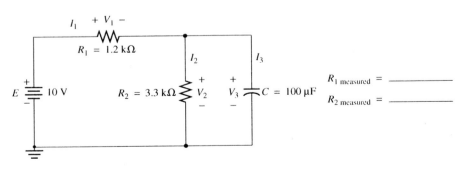

FIG. 14.3

(b) Using the measured values, calculate the theoretical steady-state levels (time greater than five time constants) of all the voltages and currents for the network of Fig. 14.3 and record in Table 14.2.

Calculation:

TABLE 14.2

	I_1	I_2	I_3	V_1	V_2	V_3
Calculated						
Measured						

(c) Energize the system and measure the voltages V_1, V_2, and V_3. Calculate the currents I_1 and I_2 from Ohm's law and the current I_3 from Kirchhoff's current law. Record all the results in Table 14.2. Compare the results with those of part 2(b).

Calculation:

Part 3   Series-Parallel *R-C* dc Network

(a) Construct the network of Fig. 14.4. Insert the measured resistor values.

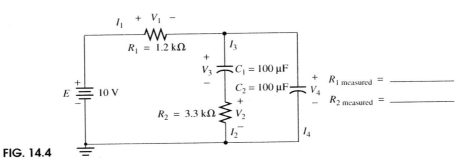

FIG. 14.4

(b) Assuming ideal capacitors and using measured resistor values, calculate the theoretical steady-state levels of the currents in Table 14.3 and voltages in Table 14.4.

Calculation:

TABLE 14.3

	I_1	I_2	I_3	I_4
Calculated				

TABLE 14.4

	V_1	V_2	V_3	V_4
Calculated				
Measured				

(c) Energize the system and measure the voltages V_1, V_2, V_3, and V_4. Record the results is the bottom row of Table 14.4. Compare the results with those in part 3(b).

Part 4   Determining C (Actual Value)

This part of the experiment will determine the actual capacitance of the capacitor. In most cases, the actual value will be more than the nameplate value.

(a) Construct the network of Fig. 14.5. Insert the measured resistance value.

$R_{measured} =$ _____

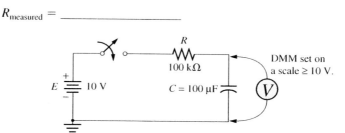

FIG. 14.5

(b) Calculate the time constant determined by the measured resistance value and the nameplate capacitance value. Record in Table 14.5.

TABLE 14.5

	Theortical	Measured
τ		

(c) Before turning on the power supply or closing the switch, be sure to discharge the capacitor by placing a lead across its terminals. Then energize the source, close the switch, and note how many seconds pass before the voltage v_C reaches 63.2% of its final value or $(0.632)(10 \text{ V}) = 6.32 \text{ V}$. Recall from the Résumé of Theory that the voltage v_C should reach 63.2% of its final steady-state value in one time constant. Record the value in Table 14.5.

(d) The actual capacitance (measured value) is then defined by

$$C_{measured} = \frac{\tau \text{ (measured)}}{R \text{ (measured)}}$$ (14.9)

Determine $C_{measured}$ for the capacitor of Fig. 14.5, recording the result in Table 14.6.

Calculation:

TABLE 14.6

	100 μF	220 μF
C_{meas}		

(e) Repeat this process and determine $C_{measured}$ for the 220-μF capacitor.

Calculation:

For the rest of this experiment, use the measured value for each capacitance.

How do the measured and nameplate values of C compare? What does the difference suggest about the actual versus nameplate levels of capacitance?

Part 5   Charging Network (Parallel Capacitors)

(a) Construct the network of Fig. 14.6. Insert the measured resistance and capacitance values.

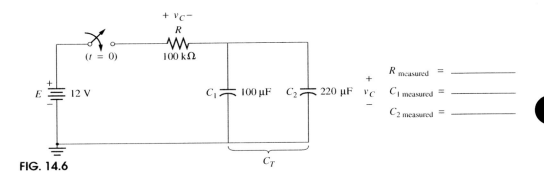

FIG. 14.6

(b) Calculate the total capacitance for the network using the measured capacitance levels. Record in Table 14.7.

Calculation:

TABLE 14.7

C_T	τ	5τ

(c) Determine the time constant for the network and record in Table 14.7.

Calculation:

(d) Calculate the charging time (5τ) for the voltage across the capacitor, C_T and record in Table 14.7.

Calculation:

(e) Using a watch, record (to the best of your ability) the voltage across the capacitor at the time intervals appearing in Table 14.8 after the switch is closed. You may want to make a test run before recording the actual levels. Complete the table using the fact that $v_R = E - v_C$. Be sure to discharge the capacitor between each run.

TABLE 14.8

$t(s)$	0	10	20	30	40	50	60	70	80	90
v_C	0 V									
v_R	12 V									

$t(s)$	100	110	120	130	140	150	160	170	180	190	200
v_C											
v_R											

(f) Plot the curves of v_C and v_R versus time on Graph 14.1. Label each curve and indicate the intervals 1τ through 5τ on the horizontal axis.

(g) What is the level of v_C after one time constant (from the graph)? Record this value in Table 14.9.

Is this level 63.2% of the final steady-state level, as dictated by the Résumé of Theory?

TABLE 14.9

	1τ	5τ	25s (equation)	25s (graph)
v_C				

(h) Record the level of v_C after five time constants in Table 14.9.

Does the level of v_C suggest that the major portion of the transient phase has passed after five time constants?

(i) Write the mathematical expression for the voltage v_C during the charging phase.

$v_C = $ _____

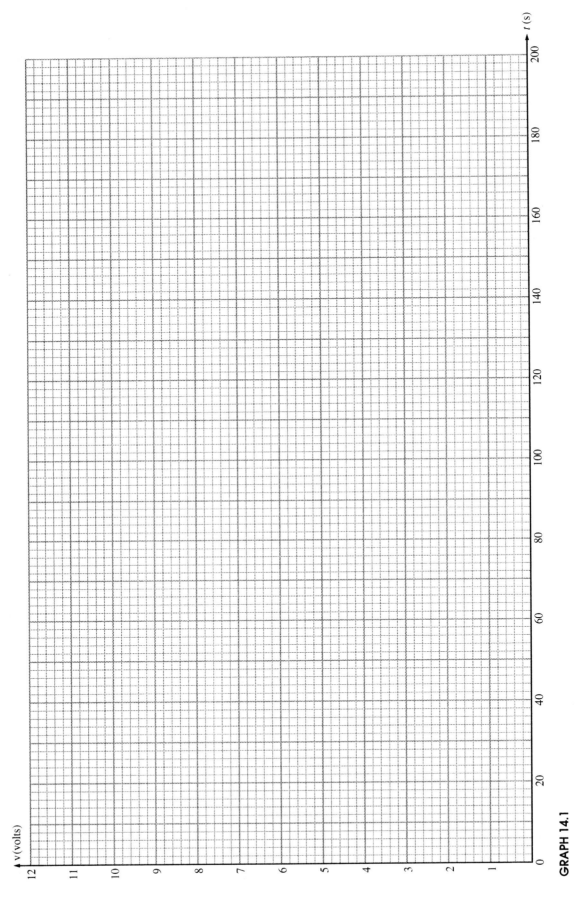

GRAPH 14.1

180

Determine the voltage v_C at $t = 25$ s by substituting the time into the preceding mathematical expression and performing the required mathematical computations. Record in Table 14.9.

Calculation:

Determine v_C at $t = 25$ s from the v_C curve of Graph 14.1 and record in Table 14.9.

How do the calculated and measured levels of v_C at $t = 25$ s compare?

Part 6 Charging Network (Series Capacitors)

(a) Construct the network of Fig. 14.7. Insert the measured resistance and capacitance values.

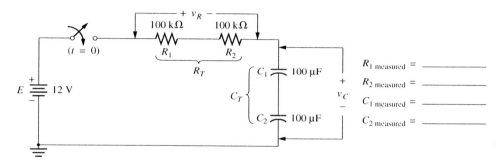

FIG. 14.7

(b) Determine the total resistance and capacitance for the network and record in Table 14.10.

Calculation:

TABLE 14.10

R_T	C_T	τ	5τ

(c) Calculate the time constant for the network and record in Table 14.10.

Calculation:

(d) Calculate the charging time (5τ) for the voltage across the capacitor, C_T and record in Table 14.10.

Calculation:

(e) Using a watch, record (to the best of your ability) the voltage across the series capacitors at the time intervals appearing in Table 14.11 after the switch is closed. You may want to make a test run before recording the actual levels. Complete the table using the fact that $v_R = E - v_C$. Again, be sure to discharge the capacitor between each run.

TABLE 14.11

t (s)	0	5	10	15	20	25	30	35
v_C	0 V							
v_R	12 V							

t (s)	40	45	50	60	70	80	90	100
v_C								
v_R								

(f) Plot the curves of v_C and v_R versus time on Graph 14.2. Label each curve and indicate the intervals 1τ through 5τ on the horizontal axis. Be sure to include the $t = 0$ s levels as determined theoretically.

(g) What is the level of v_C after one time constant (from the graph)? Record in Table 14.12.

TABLE 14.12

	1τ	5τ
v_C		

Is this level 63.2% of the final steady-state level, as dictated by the Résumé of Theory?

(h) Record the level of v_C after five time constants in Table 14.12.

Does the level of v_C suggest that the major portion of the transient phase has passed after five time constants?

(i) Write the mathematical expression for the voltage v_R during the charging phase.

$v_R = $ _____

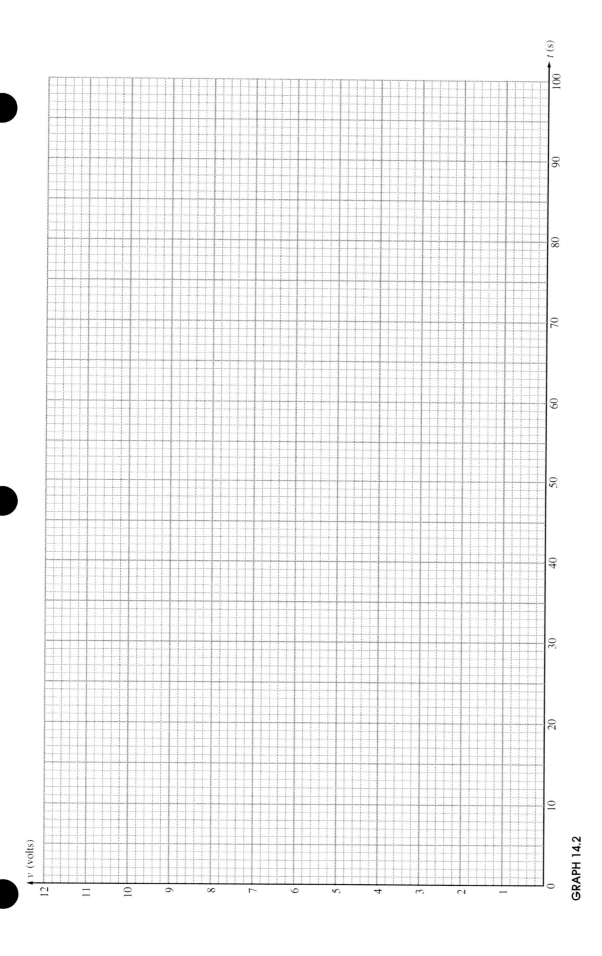

GRAPH 14.2

Using the preceding mathematical expression, determine the time t when the voltage v_R has dropped to 50% of the value at $t = 0$ s. Record in Table 14.13.

TABLE 14.13

	$t(50\%\text{-equation})$	$t(50\%\text{-graph})$
v_R		

Using the plot of v_R on Graph 14.2, determine the time t at which v_R dropped to 50% of the value at $t = 0$ s. Record in Table 14.13. How does it compare to the theoretical value just determined?

Part 7   Applying Thevenin's Theorem

(a) Construct the network of Fig. 14.8. Insert the measured values of the resistors and the capacitor.

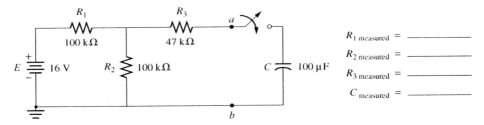

$R_{1\text{ measured}} =$ _____

$R_{2\text{ measured}} =$ _____

$R_{3\text{ measured}} =$ _____

$C_{\text{ measured}} =$ _____

FIG. 14.8

(b) Using Thevenin's theorem calculate the Thevenin resistance for the network to the left of the capacitor (between terminals a and b). Record in Table 14.14.

Calculation:

TABLE 14.14

R_{Th}	E_{Th}	1τ	5τ

Using Thevenin's theorem, calculate the open-circuit Thevenin voltage between terminals *a* and *b* for the network to the left of the capacitor. Record in Table 14.14.

(c) In the space below, redraw the network of Fig. 14.8 with the equivalent Thevenin circuit in place to the left of the switch and capacitor.

(d) Calculate the resulting time constant (τ) and charging time (5τ) for the voltage across the capacitor *C* after the switch is closed. Record in Table 14.14.

Calculation:

(e) Write the mathematical expression for the charging voltage v_C and determine the voltage v_C after one time constant. Record the result in Table 14.15.

$v_C =$ _____

Calculation:

TABLE 14.15

	1τ (equation)	1τ (measured)
v_C		

(f) Close the switch for the network of Fig. 14.8 and record the level of v_C after one time constant. Record the result in Table 14.15.

How does the value of v_C after one time constant compare with the calculated value of part 6(d)?

(g) Are the results of Parts 6(e) and 6(f) sufficiently close to validate the Thevenin equivalent circuit?

EXERCISES

1. The voltage v_C for the network of Fig. 14.9 has risen to 16 V 5 s after the switch was closed. Determine the value of C in microfarads.

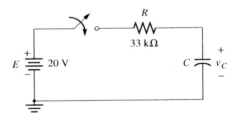

FIG. 14.9

$C =$ _____

2. Determine the mathematical expression for the voltage v_C of Fig. 14.10 following the closing of the switch.

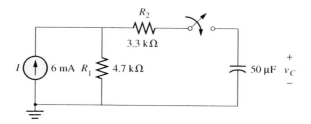

FIG. 14.10

$v_C =$ _____

3. Using PSpice or Multisim, obtain a plot of v_C for the network of Fig. 14.6 after the switch is closed. Compare the values of v_C as 1τ and 5τ using the data from Table 14.9. Enter all the results in Table 14.16.

TABLE 14.16

v_C	Table 14.9	PSpice or Multisim
1τ		
5τ		

Compare results and comment accordingly. Attach all appropriate printouts.

Name:_____

Date:_____

Course and Section:_____

Instructor:_____

R-L and *R-L-C* Circuits with a dc Source Voltage

OBJECTIVES

1. **Verify conclusions regarding the behavior of inductors in a steady-state dc network.**
2. **Note the characteristics of an inductor and capacitor in the same dc network under steady-state conditions.**
3. **Become aware of the impact of the relationship $v_L = L \, di_L/dt$ on the voltage and current levels of an *R-L* network.**

EQUIPMENT REQUIRED

Resistors

1—22-Ω, 470-Ω, 1-kΩ, 2.2-kΩ (1/4-W)

Inductors

1—10-mH

Capacitors

1—1-μF

Instruments

1—DMM
1—dc Power supply

Miscellaneous

1—SPST switch
1—Neon bulb (120 V, any color)

EQUIPMENT ISSUED

TABLE 15.0

Item	Manufacturer and Model No.	Laboratory Serial No.
DMM		
Power supply		

RÉSUMÉ OF THEORY

The inductor, like the capacitor, is an energy-storing device. The capacitor stores energy in the form of an electric field while the inductor stores it in the form of a magnetic field. The energy stored by an inductor (in joules) is given by

$$W = \frac{1}{2} LI^2 \qquad \textbf{(15.1)}$$

In any circuit containing an inductor, the voltage across the inductor is determined by the inductance (L) and the rate of change of the current through the inductor:

$$v_L = L \frac{\text{Change of Current}}{\text{Change of Time}} = L \frac{\Delta I}{\Delta t} \qquad \textbf{(15.2)}$$

If the current is constant (as in a dc circuit), $\Delta I/\Delta t = 0$, and the only voltage drop across the inductor is due to the dc resistance of the wire that makes up the inductor. Because an inductor has a dc resistance, we draw it schematically as shown in Fig. 15.1.

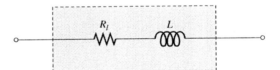

FIG. 15.1

The total inductance of inductors in series is given by

$$L_T = L_1 + L_2 + L_3 + \cdots + L_N \qquad \textbf{(15.3)}$$

and in parallel by

$$\frac{1}{L_T} = \frac{1}{L_1} + \frac{1}{L_2} + \frac{1}{L_3} + \cdots + \frac{1}{L_N} \qquad \textbf{(15.4)}$$

Note the correspondence of Eqs. (15.3) and (15.4) to the equations for resistive elements.

PROCEDURE

Part 1 Series *R-L* dc Circuit

Construct the circuit of Fig. 15.2. Insert the measured values of R_1 and R_l.

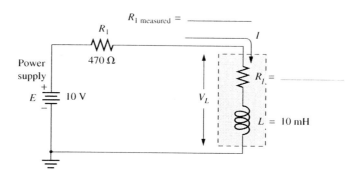

FIG. 15.2

(a) Calculate the current (I) and the voltage (V_L) for steady-state conditions. Since $5\tau = 5L/R = 5(10 \times 10^{-3})/470 \cong 0.1$ ms, steady-state conditions essentially exist once the network is constructed. Include the effects of R_l. Record the calculated values in the first column of Table 15.1.

Calculation:

TABLE 15.1

	Calculated	Measured
I		
V_L		
R_l		

(b) Measure V_L and calculate I by first measuring the voltage across the 470-Ω resistor and using Ohm's law with the measured resistor value. Record the results in column two of Table 15.1.

Calculation:

Calculate R_l using Ohm's law: $R_l = V_L/I$. Record the result in the last column of Table 15.1.

Calculation:

How does the value of R_l in part 1(b) compare with the value measured with the ohmmeter?

Is it possible for a practical inductor to have $R_l = 0\ \Omega$? Explain.

Part 2 Parallel *R-L* dc Circuit

Construct the circuit of Fig. 15.3. Insert the measured value of each resistor.

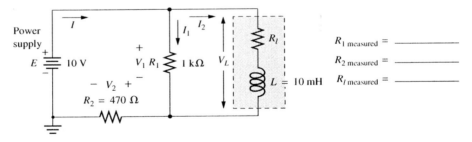

$R_{1\ \text{measured}} =$ _____

$R_{2\ \text{measured}} =$ _____

$R_{l\ \text{measured}} =$ _____

FIG. 15.3

(a) Calculate I, I_1, and I_2 assuming an ideal inductor ($R_l = 0\ \Omega$). Use measured resistor values. Record the results in Table 15.2, first column.

Calculation:

TABLE 15.2

	Calculated	Measured
I		
I_1		
I_2		

(b) Measure V_1 and V_2, and using measured resistor values calculate I, I_1, and I_2 using the equations

$$I = \frac{V_2}{R_2} \qquad I_1 = \frac{V_1}{R_1} \qquad I_2 = I - I_1$$

$V_{1_{(meas.)}} =$ _____, $V_{2_{(meas.)}} =$ _____

Record the currents in Table 15.2, second column.

Calculation:

How do the calculated (part 2a) and measured (part 2b) levels of I, I_1, and I_2 compare?

(c) Recalculate I_1 using the measured value of R_l.

Calculation:

I_1 (with R_l) = _____

How does this level of I_1 compare with the result of part 2(a) with $R_l = 0 \ \Omega$?

Based on these results, was it an appropriate approximation to assume $R_l = 0 \ \Omega$ for the initial calculations?

Part 3   Series-Parallel *R-L-C* dc Circuit

Construct the network of Fig. 15.4. Insert the measured values of all resistors (including R_l).

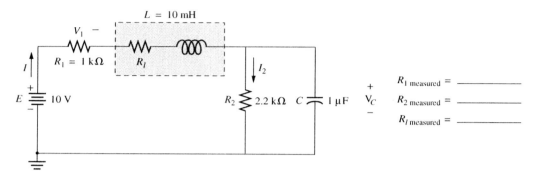

FIG. 15.4

(a) Using measured resistor values, calculate I, I_2, V_1 and V_C and record the results in Table 15.3.

Calculation:

TABLE 15.3

	V_1	V_C	I	I_2
Calculated				
Measured				

(b) Measure V_1 and V_C and using measured resistor values calculate I and I_2 from the equations

$$I = V_1/R_1 \qquad I_2 = \frac{V_2}{R_2} = \frac{V_C}{R_2}$$

$V_{1(meas.)}$ = _____, $V_{C(meas.)}$ = _____

Record the results in Table 15.3, second row.

Calculation:

Compare the measured levels with the calculated results of part 3(a).

(**c**) Calculate the energy stored by the capacitor and inductor and record below.

Calculation:

$$W_C = \underline{\hspace{3cm}}, \ W_L = \underline{\hspace{3cm}}$$

Part 4   Induced Voltage in an Inductor

(**a**) Place your power supply across the neon bulb, as shown in Fig. 15.5. The bulb is rated at 120 V for full brightness. Increase the power supply voltage to its maximum value and note whether the bulb lights up. Record the maximum value of the supply below:

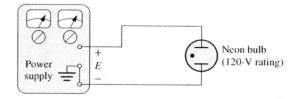

FIG. 15.5

$E_{\max} = \underline{\hspace{3cm}}$ Did the bulb light up? (Yes or No) $\underline{\hspace{3cm}}$

(**b**) Construct the network of Fig. 15.6 and set the supply to 4 V, which is 1/30 the rated voltage level of the bulb. Close the switch and measure the voltage V_1. Then calculate the current I delivered by the source using Ohm's law and the measured resistance for R_1. Record the values in Table 15.4.

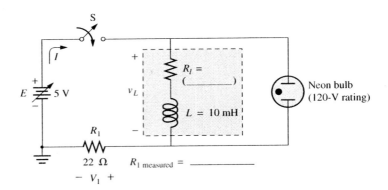

FIG. 15.6

Calculation:

TABLE 15.4

V_1	
I	
W_L	
Δt	

    Before the bulb "fires" (turns on), it can be considered an open circuit for most practical applications. The open-circuit equivalence will result in a current through the coil equal to the source current. The energy stored by the coil can then be determined using the equation $W_L = 1/2 \, LI^2$. Calculate the energy stored by the coil and record in Table 15.4.

Calculation:

Now open the switch and note the effect on the lamp. Describe in sentence form the observed effect of quickly changing the current through the coil by opening the switch.

Since the voltage across the bulb and coil is determined by $v_{bulb} = v_L = L(\Delta i_L/\Delta t)$, the change in current was sufficiently high to develop a product $L(\Delta i_L/\Delta t)$ equal to the rated voltage. Since the current dropped from the level I determined earlier to 0 A, $\Delta I = I - 0 = I$ in Eq. (15.2) and v_L must be 120 V for full brightness. Substituting, we obtain

$$v_L = 120 \text{ V} = L\frac{\Delta I}{\Delta t} = L\frac{I}{\Delta t}$$

The switching time can then be determined from

$$\Delta t = \frac{LI}{120 \text{ V}}$$

Calculate Δt for your level of I and insert the results in Table 15.4.

Calculation:

The preceding experimentation has demonstrated that the magnitude of the voltage induced across a coil is directly related to how quickly the current changes through the coil. If the current through a coil is a fixed level such as occurring for steady-state dc, the voltage across the coil is 0 V. However, if the current through the coil can be made to change very quickly, a significant voltage can be generated across the coil.

(c) Determine the time constant for the network of Fig. 15.6 during the storage phase. In the circuit resistance, include the internal resistance of the coil.

Calculation:

$\tau = $ _____

(d) On an approximate basis, how long will it take for the voltage across the coil to reach its steady-state level?

Calculation:

$5\tau =$ _____

(e) Write the mathematical expressions for the voltage v_L and current i_L during the storage phase (assuming $R_l = 0\ \Omega$).

$v_L =$ _____

$i_L =$ _____

EXERCISES

1. An automobile ignition coil (an inductor) has an inductance of 5 H. The initial current is 2 A, and $\Delta t = 0.4$ ms. Calculate the induced voltage at the spark plug.

$V_L =$ _____

2. Determine the mathematical expressions for the current i_L and voltage v_L for the network of Fig. 15.7 after the switch is closed.

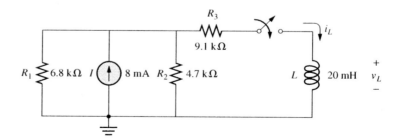

FIG. 15.7

$i_L =$ _____

$v_L =$ _____

3. Place a switch in series with the 10 V supply of Fig. 15.3. Following the closing of the switch, use PSpice or Multisim to obtain a plot of the voltage across the coil. Use mathematical techniques to determine the voltage across the coil at 1τ and 5τ and record in Table 15.5. Using the PSpice or Multisim results, determine v_L or 1τ and 5τ and insert in Table 15.5.

Calculation:

TABLE 15.5

v_L	Calculated	PSpice or Multisim
1τ		
5τ		

How do the calculated and computer results compare? Attach all appropriate printouts.

EXPERIMENT dc **16**

Design of a dc Ammeter and Voltmeter and Meter Loading Effects

OBJECTIVES

1. To design and test a dc voltmeter.
2. To design and test a dc ammeter.
3. Become aware of the importance of considering meter loading effects.

EQUIPMENT REQUIRED

Resistors

1—10-Ω, 47-Ω (1/4-W)

2—100-kΩ (1/4-W)

1—0–10-kΩ potentiometer

Instruments

1—VOM

1—DMM

1—dc Power supply

1—1-mA, 1000-Ω d'Arsonval meter movement

EQUIPMENT ISSUED

TABLE 16.0

Item	Manufacturer and Model No.	Laboratory Serial No.
VOM		
DMM		
Power supply		
Meter movement		

RÉSUMÉ OF THEORY

The basic meter movement is a current-sensitive device. Its construction consists of a coil suspended in a magnetic field with a pointer attached to it. When current is passed through the coil, an interaction of magnetic fields will cause the coil to rotate on its axis. The attached pointer will then indicate a particular angular displacement proportional to the magnitude of the current through the coil. D'Arsonval meter movements are delicate and sensitive dc milliammeters or microammeters used in the design and construction of dc voltmeters and ammeters.

A dc movement (d'Arsonval) can be used to construct a dc voltmeter by connecting a resistor in series with the movement, as shown in Fig. 16.1(a). This series resistor is called the *multiplier.* An ammeter (Fig. 16.1(b)) is constructed by placing a resistor in parallel with the meter movement. This resistor is called the *shunt* resistor.

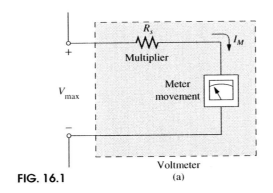

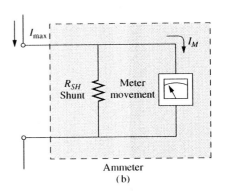

FIG. 16.1

Voltmeter (a)   Ammeter (b)

The value of R_s can be calculated as follows:

$$R_s = \frac{V_{max}}{I_M} - R_M$$

The value of R_{SH} can be calculated as follows:

$$R_{SH} = \left(\frac{I_M}{I_{max} - I_M}\right)(R_M) \qquad \textbf{(16.1)}$$

where R_s = multiplier resistor
R_{SH} = shunt resistor
I_M = full-scale deflection current of the meter movement
R_M = internal resistance of the meter movement
V_{max} = maximum value of the voltage to be measured for the range being designed
I_{max} = maximum value of the current to be measured for the range being designed

EXAMPLE

Using a 0–5-mA meter movement with an internal resistance of 5 kΩ, design

(a) a 0–50-V voltmeter

(b) a 0–100-mA milliammeter

Solutions

Given I_M = 5 mA and R_M = 5 kΩ, then

(a) $R_s = \dfrac{V_{max}}{I_M} - R_M = \dfrac{50}{5 \times 10^{-3}} - 5 \times 10^3 = 5 \text{ k}\Omega$

(note Fig. 16.2 (a)) and

(b) $R_{SH} = \left(\dfrac{I_M}{I_{max} - I_M}\right)(R_M)$

$= \left[\dfrac{5 \times 10^{-3}}{(100 \times 10^{-3}) - (5 \times 10^{-3})}\right](5 \times 10^3)$

$= 263 \text{ }\Omega$

(note Fig. 16.2(b)).

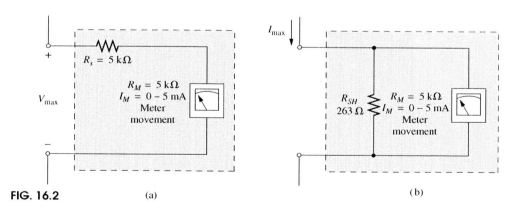

FIG. 16.2   (a)   (b)

The current sensitivity of a meter is a measure of the current necessary to obtain a full-scale deflection of the meter movement. The inverse of current sensitivity of a voltmeter is the ohm/volt sensitivity of the meter. Therefore, the dc multirange voltmeter, with a 1000-Ω/V sensitivity, has a current sensitivity of 1/1000, or 1 mA. The VOM, with a 20,000-Ω/V sensitivity for dc, has a current sensitivity of 1/20,000, or 50 μA.

Ideally, the internal resistance of a voltmeter should be infinite to minimize its impact on a circuit when taking measurements. As shown in Fig. 16.3, the internal resistance of the meter will appear in parallel with the element under investigation. The effect is to reduce the net resistance of the network, since $R_2 > R_2 \| R_{int}$, causing V_2 to decrease in magnitude. The closer R_{int} is to R_2, the more it will affect the reading. As a rule, R_{int} should be more than 10 times R_2 to ensure a reasonable level of accuracy in the reading. The internal resistance of a voltmeter is the series combination of the multiplier and movement resistances, as shown in Fig. 16.3.

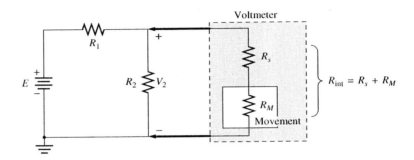

FIG. 16.3

The accuracy in percent of the full-scale reading indicates the maximum error that will occur in the meter. The Simpson 377 dc voltmeter has an accuracy of ±5% of full scale. Thus, when set on the 100-V scale, the reading may be in error by ±5 V at any point on the scale. Therefore, for maximum accuracy, always choose a scale setting that will give a reading (meter deflection) as close to full scale as possible.

PROCEDURE

Part 1   Design of a dc Voltmeter

(a) Use the 1000-Ω, 1-mA meter movement to design a 0–10-V dc voltmeter using the appropriate variable resistor. Ask your instructor to check your design before proceeding. Show all calculations and sketch the design.

$R_s =$ _____

Testing the Designed Voltmeter:

(b) Connect your newly constructed voltmeter in parallel with a voltage source and the DMM as shown in Fig. 16.4. Set the variable resistor to R_s with your DMM in the ohmmeter mode.

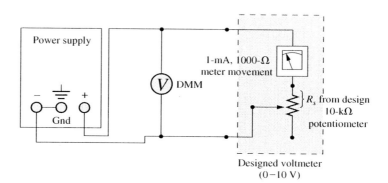

FIG. 16.4

Now vary the power supply from 0 to 10 V in 1-V steps and record your readings in Table 16.1.

TABLE 16.1

DMM	Designed Voltmeter	% Difference
1 V		
2 V		
3 V		
4 V		
5 V		
6 V		
7 V		
8 V		
9 V		
10 V		

Calculate the percent difference of the reading of your voltmeter and record in Table 16.1.

$$\% \text{ Difference} = \left| \frac{\text{DMM} - \text{Designed}}{\text{DMM}} \right| \times 100\%$$

Calculation:

What do you conclude from these results?

What is the sensitivity (Ω/V) of your voltmeter?

Part 2   Design of a dc Ammeter

Use the meter movement (1-mA, 1000-Ω) to design a 100-mA ammeter. Let your instructor check your design before proceeding. Then connect the ammeter into the circuit shown in Fig. 16.5 and compare its reading with that of the multirange dc milliammeter set on the proper scale. Start with the power supply set to zero and increase it so that the current I varies from 0 to 100 mA in 10-mA steps using the DMM in the milliammeter mode. Record the data in Table 16.2. Show your calculations, and sketch the design.

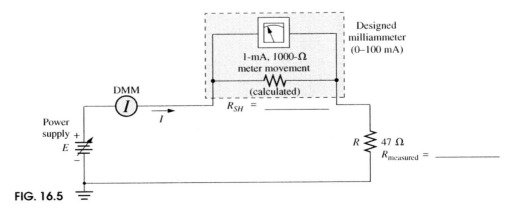

FIG. 16.5

$R_{SH} =$ _____

TABLE 16.2

DMM	Designed Milliammeter	% Difference
10 mA		
20 mA		
30 mA		
40 mA		
50 mA		
60 mA		
70 mA		
80 mA		
90 mA		
100 mA		

Calculate the percent difference of the reading of your milliammeter and record in Table 16.2.

Calculation:

What conclusions do you draw from the preceding results?

Part 3   Meter Loading Effects

(a) Construct the circuit of Fig. 16.6.

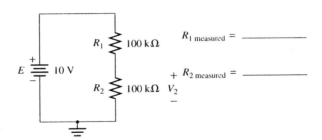

R_1 measured = _____

R_2 measured = _____

FIG. 16.6

What is the theoretical value of the voltage V_2? Use measured resistor values. Show all work! Record as the first entry in Table 16.3.

Calculation:

(b) Read the voltage V_2 using the DMM and the VOM on the 10-V scale. Record the results in columns 2 and 3 of Table 16.3.

(c) Most DMMs have an internal impedance of 10 MΩ or greater and can be considered an open circuit across the 100-kΩ resistors. The VOM, however, has an ohm/volt rating of 20,000 and therefore an internal resistance of (10 V)(20,000 Ω/V) = 200 kΩ on the 10-V scale. Using this internal resistance value for the VOM, calculate the voltage V_2 and compare with the result of part 3(b) for the VOM. Record the results in column four of Table 16.3.

Calculation:

(d) Will the 50-V scale of the VOM give a better reading? Record the measured value in the final column of Table 16.3 and note any disadvantages associated with using this higher scale.

(e) If the resistors were replaced by 1-kΩ values, would there be an improvement in the readings of the VOM? Show why with a sample calculation.

Calculation:

TABLE 16.3

	Theoretical Result	DMM	VOM	Calculated	50 V scale
V_2					

EXERCISES

1. An ammeter with an accuracy of $\pm3\%$ of full scale reads 2.1 A when set on the 5-A scale. Within what range of values (about 2.1 A) is this meter accurate?

Range = _____

2. A meter has an ohm/volt sensitivity of 100,000. What is its current sensitivity in microamperes?

CS = _____

Wheatstone Bridge and Δ-Y Conversions

OBJECTIVES

1. Become familiar with the construction and terminal characteristics of the Wheatstone Bridge.
2. Use a commercial Wheatstone bridge to measure an unknown resistance.
3. Apply a Δ-Y conversion to a Wheatstone bridge to test the Δ-Y conversion equations.

EQUIPMENT REQUIRED

Resistors

1—91-Ω (1/4-W)

2—220-Ω (1/4-W)

3—330-Ω, 1-kΩ (1/4-W)

1—0–1-kΩ potentiometer

1—Unmarked fixed resistor in the range 47 Ω to 220 Ω

Instruments

1—DMM

1—dc Power supply

1—Commercial Wheatstone bridge (if available)

EQUIPMENT ISSUED

TABLE 17.0

Item	Manufacturer and Model No.	Laboratory Serial No.
DMM		
Power supply		
Wheatstone bridge		

RÉSUMÉ OF THEORY

The Wheatstone bridge is an instrument used to make precision measurements of unknown resistance levels. The basic configuration appears in Fig. 17.1. The unknown resistance is R_x, and R_1, R_2, and R_3 are precision resistors of known value.

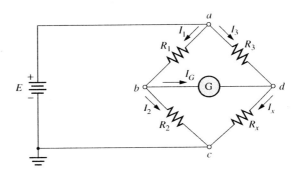

FIG. 17.1

The network is balanced when the galvanometer (G) reads zero.
We know from circuit theory that if $I_G = 0$ A, the voltage V_{bd} is zero, and

$$V_{ab} = V_{ad} \quad \text{and} \quad V_{bc} = V_{dc}$$

By substitution,

$$I_1 R_1 = I_3 R_3 \tag{17.1}$$

and

$$I_2 R_2 = I_x R_x \tag{17.2}$$

Solving Eq. (17.1) for I_1 yields

$$I_1 = \frac{I_3 R_3}{R_1}$$

Substituting I_1 for I_2 and I_3 for I_x in Eq. (17.2) (since $I_1 = I_2$ and $I_3 = I_x$ with $I_G = 0$ A), we have

$$I_1 R_2 = I_3 R_x \quad \text{or} \quad \left(\frac{I_3 R_3}{R_1} \right) R_2 = I_3 R_x$$

Canceling I_3 from both sides and solving for R_x, we obtain

$$R_x = \frac{R_2}{R_1} R_3 \tag{17.3}$$

or, in the ratio form,

$$\boxed{\frac{R_1}{R_2} = \frac{R_3}{R_x}} \tag{17.4}$$

In the commercial Wheatstone bridge, R_1 and R_2 are variable in decade steps so that the ratio R_1/R_2 is a decimal or integral multiplier. R_3 is a continuous variable resistor, such as a slide-wire rheostat. Before the unknown resistor is connected to the terminals of the commercial bridge, the R_1/R_2 ratio (called the *factor of the ratio arms*) is adjusted for that particular unknown resistor. After the resistor is connected, R_3 is adjusted until there is no detectable current indicated by the galvanometer. (Galvanometer sensitivities are usually 10^{-10} A or better.) The unknown resistance value is the ratio factor times the R_3 setting.

There are certain network configurations in which the resistors do not appear to be in series or parallel. Under these conditions, it is necessary to convert the network in question from one form to another. The two networks to be investigated in this experiment are the delta (Δ) and the wye (Y), both of which appear in Fig. 17.2. To convert a Δ to a Y (or vice versa), we use the following conversion equations:

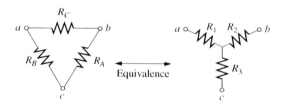

FIG. 17.2

$$\boxed{R_1 = \frac{R_B R_C}{R_A + R_B + R_C} \qquad R_2 = \frac{R_A R_C}{R_A + R_B + R_C} \qquad R_3 = \frac{R_A R_B}{R_A + R_B + R_C}} \tag{17.5}$$

$$\boxed{R_A = \frac{R_1 R_2 + R_1 R_3 + R_2 R_3}{R_1} \qquad R_B = \frac{R_1 R_2 + R_1 R_3 + R_2 R_3}{R_2} \\ R_C = \frac{R_1 R_2 + R_1 R_3 + R_2 R_3}{R_3}} \tag{17.6}$$

If $R_A = R_B = R_C$,

$$\boxed{R_Y = \frac{R_\Delta}{3}} \tag{17.7}$$

If $R_1 = R_2 = R_3$,

$$\boxed{R_\Delta = 3R_Y} \tag{17.8}$$

PROCEDURE

Part 1   Wheatstone Bridge Network

(a) Construct the network of Fig. 17.3. Insert the measured value of each resistor and set the potentiometer to the maximum resistance setting.

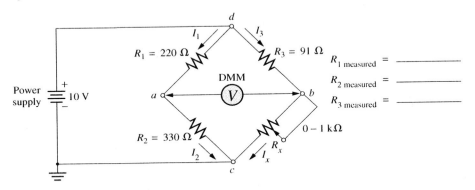

$R_{1\ measured}$ = _____

$R_{2\ measured}$ = _____

$R_{3\ measured}$ = _____

FIG. 17.3

(b) Starting with the meter on a higher voltage scale, vary the potentiometer until the voltage V_{ab} is as close to zero as possible. Then drop the voltage scales to the lowest range possible to set the voltage V_{ab} as close to 0 V as possible. The bridge is now balanced.

(c) Measure the voltages V_{da}, V_{db}, V_{ac}, and V_{bc} and record in Table 17.1.

TABLE 17.1

V_{da}	V_{db}	V_{ac}	V_{bc}

(d) Calculate the currents I_1 and I_2 using Ohm's law and record in Table 17.3. Are they equal, as defined in the Résumé of Theory?

Calculation:

(e) Disconnect one lead of the potentiometer (used as a rheostat) and measure its resistance. Record the value in Table 17.2.

TABLE 17.2

R_{pot} (e)	R_{pot} (g)	R_3 (calc.)	R_3 (meas.)	R_{max}	R_3 (bridge)

Calculate the currents I_3 and I_x using the results of part 1(c) and Ohm's law and record in Table 17.3. Are they equal, as defined in the Résumé of Theory?

Calculation:

TABLE 17.3

I_1	I_2	I_3	I_x

Calculation:

(f) Verify that the following ratio is satisfied:

$$\frac{R_1}{R_2} = \frac{R_3}{R_x}$$

(g) Replace the 91-Ω resistor (R_3) by the unknown resistor. Proceed as before to adjust the potentiometer until $V_{ab} \cong 0$ V. Remove the variable resistor and measure its resistance with the ohmmeter section of your multimeter. Record in column two of Table 17.2.

(h) Calculate the unknown resistance R_3 using Eq. (17.4) and record in column three of Table 17.2.

Calculation:

   (i) Measure the unknown resistance R_3 with the ohmmeter section of the DMM and compare to the result of part 1(h). Record the measured value in column four of Table 17.2.

   (j) What is the maximum value of resistance that this network could measure by placing the unknown in the R_3 position? Record the determined value in column five of Table 17.2.

Calculation:

Part 2   Commercial Wheatstone Bridge

Use the commercial Wheatstone bridge to measure the resistance of the unknown resistor and record in the final column of Table 17.2.

   Compare with the values found in parts 1(h) and 1(i).

Part 3   Δ-Y Conversions

   (a) Construct the network of Fig. 17.4. Insert the measured values of the resistors but assume for the moment that each 1-kΩ resistor is exactly 1 kΩ.

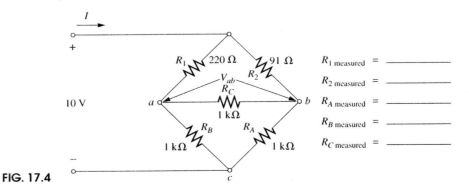

FIG. 17.4

(b) Calculate the source current I and the voltage V_{ab} using any method other than a Δ-Y conversion (that is, mesh analysis, nodal analysis, and so on). Record the results in column one of Table 17.4. Use $R_A = R_B = R_C = 1$ kΩ (exactly).

Calculation:

TABLE 17.4

	Calculated	Measured	Meas. (equivalent)
I			
V_{ab}			

(c) Measure the current I and the voltage V_{ab} and compare to the results of part 3(b). Record the measurements in column two of Table 17.4.

(d) Calculate the equivalent Y for the Δ formed by the three 1-kΩ resistors (assume all are exactly 1 kΩ). Draw the equivalent circuit with the Δ replaced by the Y. Insert the measured values of the resistors chosen in Fig. 17.5.

Calculation:

(e) Construct the network of Fig. 17.5 and measure the current I and the voltage V_{ab}. Record the measurements in column three of Table 17.4. Are they approximately the same as those obtained in part 3(c)? Has the conversion process been verified? Try to account for any major differences.

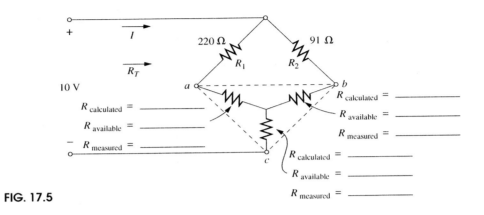

FIG. 17.5

(f) Calculate the input resistance to the network of part 3(d) using the measured resistor values. Record in column one of Table 17.5.

TABLE 17.5

	Calculated (3f)	Measured (3g)	Calculated (3h)
R_T			

(g) Disconnect the supply and measure the input resistance to the network of part 3(e) using the ohmmeter section of the DMM. Record the measurement in column two of Table 17.5. Compare with the results of part 3(f).

(**h**) Determine the input resistance to the network of part 3(a) using $R_T = E/I$ and compare with the results of part 3(g). Record the result in the last column of Table 17.5. Should they compare? Why?

Calculation:

(**i**) Calculate I using the result of part 3(f) and compare with the value measured in part 3(e).

Calculation:

$$I = \underline{\hspace{2cm}}$$

EXERCISES

1. Perform a Δ-Y conversion on the upper delta of Fig. 16.4 and calculate the current I with 10 V applied at the input.

$$I = \underline{\hspace{2cm}}$$

How does this level of I compare with the calculated and measured values of the experiment?

2. Referring to Fig. 17.1, if R_1 and R_2 are 1.5 Ω and 3.3 Ω, respectively, can the bridge be balanced by resistors of 1.5 MΩ and 3.3 MΩ for R_3 and R_x, respectively? Explain your answer.

3. Using PSpice or Multisim, determine the current I and the voltage V_{ab} for the network of Fig. 17.4. In Table 17.6, record the value of each obtained in part 3(c) and through computer methods.

TABLE 17.6

	Measured (3c)	PSpice or Multisim
I		
V_{ab}		

Compare the results and comment accordingly. Attach all appropriate printouts.

Ohmmeter Circuits

OBJECTIVES

1. Design and test a series ohmmeter circuit.
2. Design and test a shunt ohmmeter circuit.
3. Examine and test a voltage divider ohmmeter configuration.

EQUIPMENT REQUIRED

Resistors (fixed)

1—1-kΩ, 22-kΩ (1/4-W)
2—10-kΩ (1/4-W)

Resistors (variable)

1—0–10-kΩ potentiometer
1—0–250-kΩ potentiometer

Instruments

1—DMM
1—dc Power supply
1—0–1-mA, 1000-Ω meter movement

EQUIPMENT ISSUED

TABLE 18.0

Item	Manufacturer and Model No.	Laboratory Serial No.
DMM		
Power supply		
Meter movement		

RÉSUMÉ OF THEORY

The ohmmeter is an instrument that is used to measure resistance. The majority of commercial ohmmeters fall into one of the following three categories:

1. Series type, Fig. 18.1(a)
2. Shunt type, Fig. 18.1(b)
3. Voltage divider type, Fig. 18.1(c)

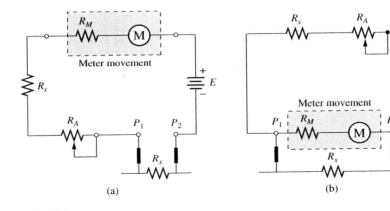
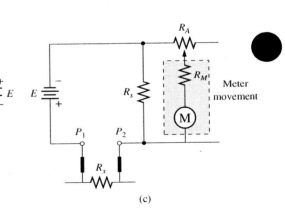

(a)   (b)   (c)

FIG. 18.1

For each configuration, the variable resistor R_A is a fine-adjust on R_s to establish full-scale deflection (I = rated maximum movement current) with a fixed supply (E).

The *series* type reads right to left. It cannot be used for low-resistance measurements (below 1 Ω).

The *shunt* type reads left to right. It is used primarily for low-resistance measurements (below 1000 Ω).

The *voltage divider* type reads right to left. It is the most versatile and most often used.

When using an ohmmeter, one must remember two important rules:

1. Never leave the ohmmeter on or, in the case of a VOM, in the ohmmeter position. The internal voltage source is always in the circuit so that the life span of the battery is shortened.
2. Always adjust the "ohms-adjust" after changing the range.

PROCEDURE

Part 1   Series Ohmmeter Circuit

Construct the circuit of Fig. 18.2

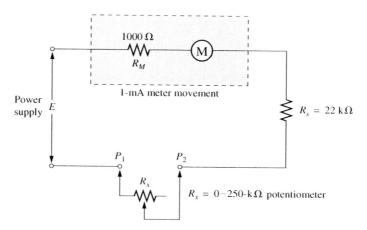

FIG. 18.2

Short the probes points P_1 and P_2 together and adjust the power supply until the meter movement reads 1 mA. Hold the probes apart (open-circuit) and read the meter movement. Record in the last row of Table 18.1.

For the remaining steps of the experiment, proceed as follows.

Use the DMM to set the potentiometer resistance to the values shown in Table 18.1. Then insert between P_1 and P_2 and read the meter (M). Record in Table 18.1.

Plot the data of Table 18.1 using the vertical and horizontal divisions in Graph 18.1.

TABLE 18.1

Resistance Across Probes	$I_{meter\ movement}$
0 kΩ	1 mA
20 kΩ	
40 kΩ	
60 kΩ	
80 kΩ	
100 kΩ	
120 kΩ	
140 kΩ	
160 kΩ	
180 kΩ	
200 kΩ	
Open circuit	

The graph is now used as a calibration chart for the ohmmeter circuit of Fig. 18.2. That is, for any unknown applied resistor, the resulting current will define an intersection with the plot of Graph 18.1 that will then define the resistance level on the horizontal axis.

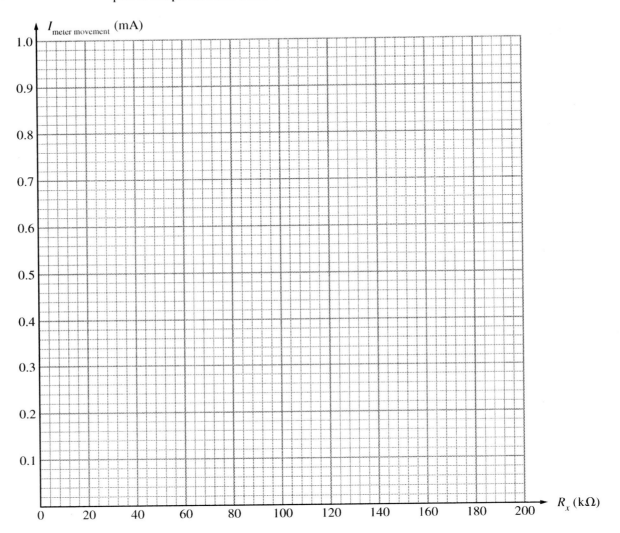

GRAPH 18.1

Vary the 0–250-kΩ potentiometer randomly for three settings and record the resulting value of I_M in column one of Table 18.2. Then use the calibration curve to determine the unknown resistance and record in column two of Table 18.2. Finally, measure the unknown resistance with the DMM and record in column three of Table 18.2.

TABLE 18.2

	I_M	$R_{(curve)}$	$R_{(DMM)}$
1			
2			
3			

Draw a meter scale for your ohmmeter; that is, complete the resistance scale of Fig. 18.3. Would the scale of this ohmmeter be considered linear or nonlinear?

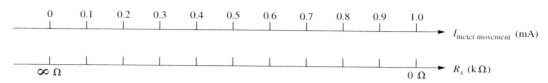

FIG. 18.3

What is the range of resistance values that this ohmmeter is capable of measuring with reasonably accurate results? Explain.

Why would you recommend that, for the greatest accuracy, the meter be read to the right of the center portion of the scale?

Part 2   Shunt Ohmmeter Circuit

Construct the circuit of Fig. 18.4

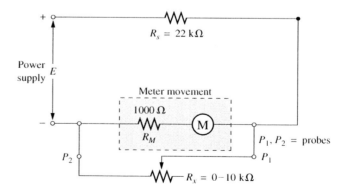

FIG. 18.4

Keeping the probes apart, adjust the power supply until the meter movement reads 1 mA. Record in Table 18.3. Short the probes together (0 Ω); then read the meter. Record in Table 18.3.

Use the DMM to set the potentiometer resistance to the values shown in Table 18.3. Then insert between P_1 and P_2 and read the meter. Record in Table 18.3. Plot the data of Table 18.3 using the horizontal and vertical divisions in Graph 18.2.

TABLE 18.3

Resistance Across Probes	$I_{\text{meter movement}}$
0 Ω	
200 Ω	
400 Ω	
800 Ω	
1,000 Ω	
2,000 Ω	
4,000 Ω	
6,000 Ω	
8,000 Ω	
10,000 Ω	

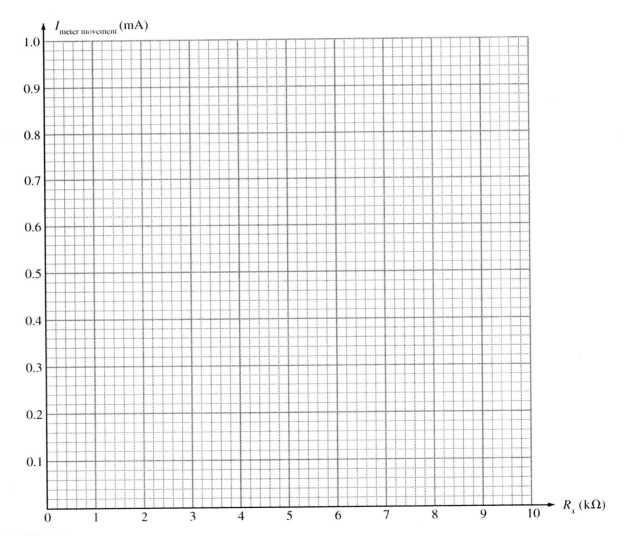

GRAPH 18.2

Graph 18.2 now can be used as a calibration chart for the shunt circuit of Fig. 18.4.

Now set the 0–10-kΩ potentiometer (R_x) to three random values and record the resulting value of I_M in column one of Table 18.4. Then use the calibration curve to determine the unknown resistance and record in column two of Table 18.4. Finally, measure the unknown resistance with the DMM and record in column three of Table 18.4.

TABLE 18.4

	I_M	$R_{(curve)}$	$R_{(DMM)}$
1			
2			
3			

How do the results compare?

Compare and contrast the two graphs of the series and shunt ohmmeters. Comment accordingly.

Draw a meter scale for your ohmmeter. That is, complete the resistance scale of Fig. 18.5. Is the scale linear or nonlinear?

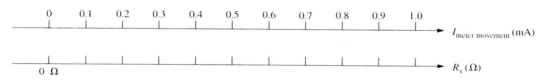

FIG. 18.5

Part 3   Voltage Divider Ohmmeter

Construct the circuit of Fig. 18.6.

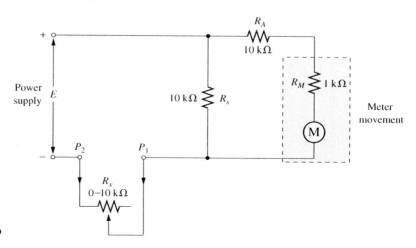

FIG. 18.6

Short the probes (0 Ω) and adjust the power supply until the meter movement reads 1 mA.
Separate the probes and read the meter and record in the last row of Table 18.5.
Record in Table 18.5. Use the DMM to set the potentiometer resistance to the values shown in Table 18.5. Read the meter movement for each resistance setting. Record in Table 18.5.

TABLE 18.5

Resistance Across Probes	$I_{meter\ movement}$
0 Ω	1 mA
500 Ω	
1,000 Ω	
2,000 Ω	
3,000 Ω	
4,000 Ω	
5,000 Ω	
8,000 Ω	
10,000 Ω	
∞	

Plot the data of Table 18.5 using the vertical and horizontal divisions in Graph 18.3.
This graph can now be used as a calibration chart for the voltage divider ohmmeter circuit of Fig. 18.6. Now set the 0–10-kΩ potentiometer (R_x) to three random values and record the resulting value of I_M in the first column of Table 18.6. Then use the calibration curve to determine the unknown resistance and record the results in column two of Table 18.6. Finally, measure the unknown resistance with the DMM and record in the last column of Table 18.6.

TABLE 18.6

	I_M	$R_{(curve)}$	$R_{(DMM)}$
1			
2			
3			

How do the results compare?

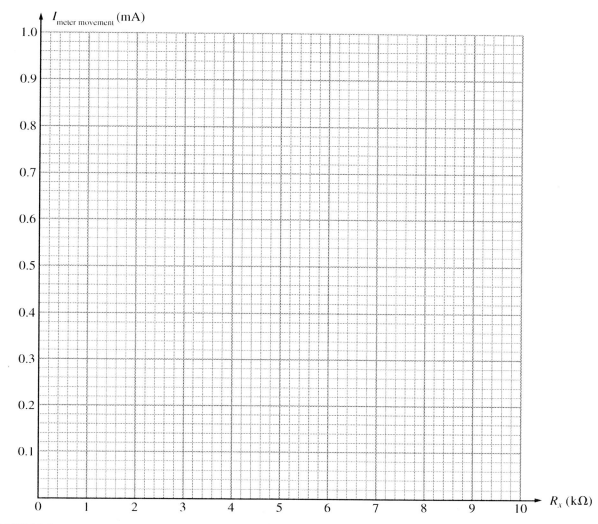

GRAPH 18.3

Draw a meter scale for your ohmmeter for the range plotted. That is, complete the resistance scale of Fig. 18.7.

FIG. 18.7

Is the scale linear or nonlinear?

EXERCISE

Summarize by comparing and contrasting the three circuits as to their advantages, disadvantages, and characteristics.

Name: _____

Date: _____

Course and Section: _____

Instructor: _____

Math Review and Calculator Fundamentals ac

MATH REVIEW

Pythagorean Theorem

Polar and Rectangular Forms of Vectors

Conversion Between Vector Forms

Addition and Subtraction of Vectors

Multiplication and Division of Vectors

CALCULATOR FUNDAMENTALS

MATH REVIEW

The analysis of sinusoidal ac networks will require the extended use of the Pythagorean theorem, trigonometric functions, and vector operations. This review will begin to establish the mathematical foundation necessary to work with each of these functions and operations in a confident, correct, and accurate manner.

It is assumed that a calculator or table is available to determine (with reasonable accuracy) the value of the trigonometric functions (sine, cosine, tangent) and the square and square root of a number.

The use of the calculator is covered later in this laboratory exercise.

Pythagorean Theorem

The Pythagorean theorem states that the square of the hypotenuse of a right triangle is equal to the sum of the squares of the other sides of the triangle. With the notation in Fig. 1.1, the theorem has the following form:

$$Z^2 = X^2 + Y^2 \qquad \text{(1.1)}$$

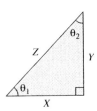

FIG. 1.1

For θ_1,

$$\sin \theta_1 = \frac{\text{Opposite}}{\text{Hypotenuse}} = \frac{Y}{Z} \qquad \text{(1.2)}$$

$$\cos \theta_1 = \frac{\text{Adjacent}}{\text{Hypotenuse}} = \frac{X}{Z} \qquad \text{(1.3)}$$

$$\tan \theta_1 = \frac{\sin \theta_1}{\cos \theta_1} = \frac{Y/Z}{X/Z} = \frac{Y}{X} = \frac{\text{Opposite}}{\text{Adjacent}} \qquad \text{(1.4)}$$

For θ_2,

$$\sin \theta_2 = \frac{X}{Z}$$

$$\cos \theta_2 = \frac{Y}{Z}$$

$$\tan \theta_2 = \frac{X}{Y}$$

To determine θ_1 and θ_2, we use

$$\theta_1 = \sin^{-1}\frac{Y}{Z} = \cos^{-1}\frac{X}{Z} = \tan^{-1}\frac{Y}{X}$$

$$\theta_2 = \sin^{-1}\frac{X}{Z} = \cos^{-1}\frac{Y}{Z} = \tan^{-1}\frac{X}{Y}$$

$$(1.5)$$

EXAMPLE 1   Find the hypotenuse Z and the angle θ_1 for the right triangle of Fig. 1.2.

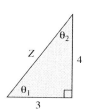

FIG. 1.2

Solution

$$Z = \sqrt{X^2 + Y^2} = \sqrt{(3)^2 + (4)^2} = \sqrt{25} = \mathbf{5}$$

and

$$\theta_1 = \tan^{-1}\frac{4}{3} = \tan^{-1}1.333\ldots \cong \mathbf{53.13°}$$

or

$$\theta_1 = \sin^{-1}\frac{4}{5} = \sin^{-1}0.8 \cong \mathbf{53.13°}$$

or

$$\theta_1 = \cos^{-1}\frac{3}{5} = \cos^{-1}0.6 \cong \mathbf{53.13°}$$

EXAMPLE 2   Determine Y and θ_1 for the right triangle of Fig. 1.3.

FIG. 1.3

Solution

$$Z^2 = X^2 + Y^2$$

or

$$Y^2 = Z^2 - X^2$$

and

$$Y = \sqrt{Z^2 - X^2}$$ **(1.6)**

Substituting,

$$Y = \sqrt{(12)^2 - (6)^2} = \sqrt{108} \cong \mathbf{10.39}$$

and

$$\theta_1 = \cos^{-1}\frac{6}{12} = \cos^{-1}0.5 = \mathbf{60°}$$

EXERCISES

1. Determine θ_2 for Fig. 1.2.

$\theta_2 = $ _____

2. Determine θ_2 for Fig. 1.3.

$\theta_2 = $ _____

3. **(a)** Determine X for the right triangle of Fig. 1.4.

$X = $ _____

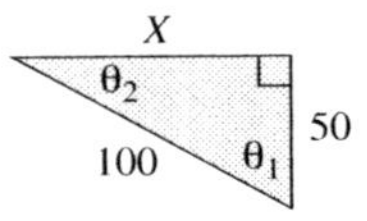

FIG. 1.4

(b) Determine θ_1 and θ_2 for Fig. 1.4.

$\theta_1 = $ _____ , $\theta_2 = $ _____

4. Refer to Fig. 1.5.

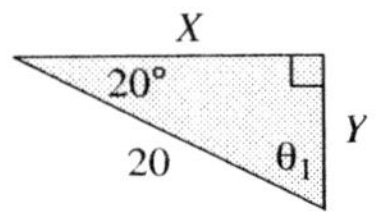

FIG. 1.5

(a) Determine Y.

$Y = $ _____

(b) Find X.

$X =$ _____

(c) Determine θ_1.

$\theta_1 =$ _____

Vector Representation

There are two common forms for representing vectors with the tail fixed at the origin. The *rectangular* form has the appearance shown in Fig. 1.6, where the real and imaginary components define the location of the head of the vector.

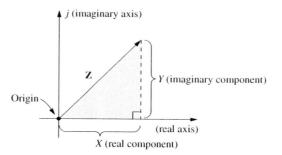

FIG. 1.6

The imaginary component is defined by the component of the vector that has the j associated with it, as appearing in the following representation of the rectangular form:

Rectangular Form:

$$\boxed{\mathbf{Z} = X + jY} \tag{1.7}$$

In the first and fourth quadrants, X is positive, whereas in the second and third quadrants, it is negative. Similarly, the imaginary component is positive in the first and second quadrants and negative in the third and fourth.

The *polar* form has the appearance shown in Fig. 1.7, where the angle θ is *always* measured from the positive real axis. The equation form is as follows:

Polar Form:

$$\boxed{\mathbf{Z} = Z \angle \theta} \tag{1.8}$$

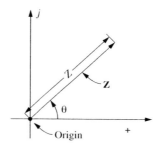

FIG. 1.7

Through Fig. 1.8, it should be fairly obvious that the two forms are related by the following equations.

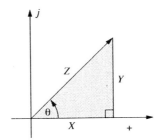

FIG. 1.8

Rectangular to Polar:

$$Z = \sqrt{X^2 + Y^2} \qquad (1.9)$$

$$\theta = \tan^{-1}\frac{Y}{X} \qquad (1.10)$$

Polar to Rectangular:

$$X = Z\cos\theta \qquad (1.11)$$

$$Y = Z\sin\theta \qquad (1.12)$$

EXAMPLE 3   Determine the rectangular and polar forms of the vector of Fig. 1.9.

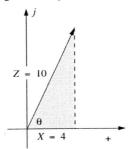

FIG. 1.9

Solution

Rectangular Form

$$Y = \sqrt{Z^2 - X^2} = \sqrt{(10)^2 - (4)^2} = \sqrt{84} \cong 9.165$$

and

$$\mathbf{Z} = \mathbf{4} + j\mathbf{9.165}$$

Polar Form

$$\theta = \tan^{-1}\frac{Y}{X} = \tan^{-1}\frac{9.165}{4} = \tan^{-1}2.291 \cong 66.42°$$

and

$$\mathbf{Z} = \mathbf{10} \angle \mathbf{66.42°}$$

EXAMPLE 4 Determine the polar form of the vector of Fig. 1.10.

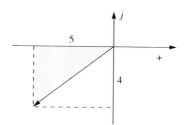

FIG. 1.10

Solution

In conversions of this type, it is usually best to isolate a right triangle, as shown in Fig. 1.11, and determine θ' and Z.

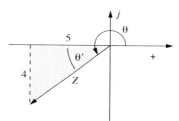

FIG. 1.11

The angle θ' is not the angle θ of the polar form, but it can be determined from the right triangle and then used to calculate θ.

In this case,

$$\mathbf{Z} = \sqrt{X^2 + Y^2} = \sqrt{(5)^2 + (4)^2} = \sqrt{41} \cong 6.4$$

and

$$\theta' = \tan^{-1} \frac{4}{5} = \tan^{-1} 0.8 \cong 38.66°$$

Using Fig. 1.11, we find

$$\theta = 180° + \theta' = 180° + 38.66° = 218.66°$$

and

$$\mathbf{Z} = 6.4 \angle 218.66°$$

EXERCISES

5. Determine the polar form of each of the following vectors:

   (a) $\mathbf{Z} = 10 + j10 =$ _____

   (b) $\mathbf{Z} = -2 + j8 =$ _____

(c) $\mathbf{Z} = -50 - j200 = $ _____

(d) $\mathbf{Z} = +0.4 - j0.8 = $ _____

6. Determine the rectangular form of each of the following vectors:

(a) $\mathbf{Z} = 6 \angle 37.5° = $ _____

(b) $\mathbf{Z} = 2 \times 10^{-3} \angle 100° = $ _____

(c) $\mathbf{Z} = 52 \angle -120° = $ _____

(d) $\mathbf{Z} = 1.8 \angle -30° = $ _____

Addition and Subtraction of Vectors

The sum or difference of vectors is normally found in the rectangular form. The operation can be performed in the polar form only if the vectors have the same angle or are out of phase by 180°.

Addition:

Rectangular Form

$$(X_1 + jY_1) + (X_2 + jY_2) = (X_1 + X_2) + j(Y_1 + Y_2) \qquad \textbf{(1.13)}$$

Polar Form

$$Z_1 \angle \theta + Z_2 \angle \theta = (Z_1 + Z_2) \angle \theta \qquad \textbf{(1.14)}$$

$$Z_1 \angle \theta + Z_2 \angle \theta \pm 180° = Z_1 \angle \theta - Z_2 \angle \theta = (Z_1 - Z_2) \angle \theta \qquad \textbf{(1.15)}$$

Subtraction:

Rectangular Form

$$(X_1 + jY_1) - (X_2 + jY_2) = (X_1 - X_2) + j(Y_1 - Y_2) \qquad \textbf{(1.16)}$$

Polar Form

$$Z_1 \angle \theta - Z_2 \angle \theta = (Z_1 - Z_2) \angle \theta \qquad \textbf{(1.17)}$$

$$Z_1 \angle \theta - Z_2 \angle \theta \pm 180° = Z_1 \angle \theta + Z_2 \angle \theta = (Z_1 + Z_2) \angle \theta \qquad \textbf{(1.18)}$$

EXAMPLE 5 Determine the sum of the following vectors:

$$\mathbf{X}_1 = 4 + j4 \qquad \mathbf{X}_2 = 6 \angle 120°$$

Solution

Using Fig. 1.12 to convert $\mathbf{X}_2$ to rectangular form,

$$Y = 6 \sin 60° = 6(0.866) = 5.196$$
$$X = 6 \cos 60° = 6(0.5) = 3$$

and

$$\mathbf{Z} = 6 \angle 120° = -3 + j5.196$$

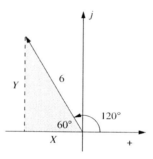

FIG. 1.12

Note in this conversion that the proper signs for X and Y are simply determined by the location of the vector.

$$\mathbf{X}_{\text{sum}} = \mathbf{X}_1 + \mathbf{X}_2 = (4 + j4) + (-3 + j5.196)$$

The parentheses around each vector identify each vector and the sign of each term.

$$\mathbf{X}_{\text{sum}} = (4 - 3) + j(4 + 5.196) = \mathbf{1 + j9.196}$$

EXAMPLE 6 Determine the difference between the following vectors:

$$\mathbf{X}_1 = -5 - j4 \qquad \mathbf{X}_2 = 3 - j8$$

Solution

$$\mathbf{X}_{\text{difference}} = \mathbf{X}_1 - \mathbf{X}_2 = (-5 - j4) - (3 - j8) = (-5 - 3) + j(-4 + 8)$$

Carefully review the sign of each term of the above equation and note how the parentheses were an aid in generating the proper signs for the following solution:

$$\mathbf{X}_{\text{difference}} = \mathbf{-8 + j4}$$

EXERCISES

7. Perform the following vector operations:

(a) $(10+j30) + (-4 - j8) = $ _____

(b) $(10 \angle 45°) - (7.071 - j7.071) = $ _____

(c) $(50 \angle 60°) + (0.8 \angle 60°) = $ _____

(d) $(16 \angle -20°) + (14 \angle 160°) = $ _____

(e) $(5000 + j1000) - (2500 \angle -60°) = $ _____

(f) $(5 \angle 0°) + (20 \angle -90°) - (6 \angle 180°) = $ _____

Multiplication and Division of Vectors

Multiplication:

Multiplication of vectors is routinely accomplished in either form, although the polar form is more direct.

Polar Form

$$ (Z_1 \angle \theta_1)(Z_2 \angle \theta_2) = Z_1 Z_2 \underline{/\theta_1 + \theta_2} $$

(1.19)

Rectangular Form

There is a lengthy equation for multiplication in the rectangular form but it seems more logical to simply state that the operation is carried out by simply multiplying the real and imaginary part of one vector by both components of the other vector.

EXAMPLE 7 Determine the product of the following vectors:

$$ \mathbf{X}_1 = 10 \angle 50° \qquad \mathbf{X}_2 = 0.4 \angle -60° $$

Solution

$$ \mathbf{X}_{\text{product}} = \mathbf{X}_1 \cdot \mathbf{X}_2 = (10 \angle 50°)(0.4 \angle -60°) = (10)(0.4) \underline{/50° + (-60°)} = \mathbf{4 \angle -10°} $$

EXAMPLE 8 Calculate the product of the following vectors:

$$ \mathbf{X}_1 = 4 + j6 \qquad \mathbf{X}_2 = 0.2 - j0.6 $$

Solution

$$
\begin{array}{r}
(4 + j6) \\
\times (0.2 - j0.6) \\
\hline
-j2.4 - j^2 3.6 \\
+0.8 + j1.2 \\
\hline
\end{array}
$$

(multiplying both terms of $\mathbf{X}_1$ by $-j0.6$)

(multiplying both terms of $\mathbf{X}_1$ by 0.2)

Since $j^2 = -1$, the sum of the terms is the following:

$$0.8 + j(1.2 - 2.4) - (-1)3.6 = 0.8 + 3.6 - j1.2$$

and

$$\mathbf{X}_{product} = \mathbf{4.4 - j1.2}$$

Division:

Although division of vectors is performed more directly in the polar form, the operation can also be performed in the rectangular form.

Polar Form

$$\boxed{\frac{Z_1 \angle \theta_1}{Z_2 \angle \theta_2} = \frac{Z_1}{Z_2} \underline{/\theta_1 - \theta_2}} \tag{1.20}$$

Rectangular Form

Division in this form is best performed by first using the following useful equation derived by multiplying the vector in the denominator by its complex conjugate (change the sign of the imaginary part):

$$\boxed{\frac{1}{X + jY} = \frac{1}{X + jY}\frac{(X - jY)}{(X - jY)} = \frac{X - jY}{X^2 + Y^2} = \frac{X}{X^2 + Y^2} - j\frac{Y}{X^2 + Y^2}} \tag{1.21}$$

Note that the result has the complex conjugate of the original vector in the numerator and a divisor equal to the sum of the squares of the two components of the rectangular form. Division then becomes a multiplication process, as follows:

$$\frac{X_1 + jY_1}{X_2 + jY_2} = (X_1 + jY_1)\left(\frac{1}{X_2 + jY_2}\right) = (X_1 + jY_1)\left(\frac{X_2}{X_2^2 + Y_2^2} - j\frac{Y_2}{X_2^2 + Y_2^2}\right)$$

EXAMPLE 9   Perform the following operations:

(a) $\dfrac{\mathbf{X}_1}{\mathbf{X}_2}$ if $\mathbf{X}_1 = 80 \angle 10°$ and $\mathbf{X}_2 = 40 \angle -20°$

(b) $\dfrac{4 - j9}{3 + j4}$

Solution

(a) $\mathbf{X}_{quotient} = \dfrac{\mathbf{X}_1}{\mathbf{X}_2} = \dfrac{80 \angle 10°}{40 \angle -20°} = \dfrac{80}{40}\underline{/10 - (-20°)} = 2\underline{/10 + 20°}$

$\qquad = \mathbf{2 \angle 30°}$

Note the need to bracket the angle from the denominator to be sure the proper sign is obtained.

(b) $\dfrac{4 - j9}{3 + j4} = (4 - j9)\left(\dfrac{3}{3^2 + 4^2} - j\dfrac{4}{3^2 + 4^2}\right) = (4 - j9)\left(\dfrac{3}{25} - j\dfrac{4}{25}\right)$

$$= \dfrac{12}{25} - j\dfrac{16}{25} - j\dfrac{27}{25} + j^2\dfrac{36}{25}$$

$$= \left(\dfrac{12}{25} - \dfrac{36}{25}\right) + j\left(\dfrac{-16}{25} - \dfrac{27}{25}\right)$$

$$= -\dfrac{24}{25} - j\dfrac{43}{25}$$

This solution eliminates the need to convert to the polar form before the division takes place.

EXERCISES

8. Perform the following operations:

    (a) $(50 \angle 60°)(4 \angle 80°) = $ _____

    (b) $(0.8 \angle -10°)(3 \angle 30°)(-4 \angle 20°) = $ _____

    (c) $(5 + j8)(4 - j2) = $ _____

    (d) $(48 \angle 20°)(10 - j1) = $ _____

    (e) $(0.9 - j0.5)(40 + j10) = $ _____

    (f) $(4000 \angle 0°)(2000 \angle -20°)(1000 + j1000) = $ _____

9. Perform the following operations:

    (a) $\dfrac{1000 \angle 60°}{20 \angle -50°} = $ _____

    (b) $\dfrac{0.008 \angle 50°}{5 \times 10^{-6} \angle 100°} = $ _____

    (c) $\dfrac{6 + j6}{2 + j2} = $ _____

    (d) $\dfrac{10 \angle 30°}{4 + j10} = $ _____

    (e) $\dfrac{20{,}000 + j10{,}500}{5000 \angle -50°} = $ _____

    (f) $\dfrac{(50 \angle 20°)(20 \angle -30°)}{(5 + j5)(4 \angle -10°)} = $ _____

CALCULATOR FUNDAMENTALS

The basic operation of the calculator and the priority it assigns to each mathematical operation were covered in detail in Experiment dc 1. In this section, we concentrate on its use with complex numbers.

Basic Operations

First let us examine the basic operations associated with some of the equations introduced in the first part of this laboratory exercise.

Pythagorean Theorem:

$$Z^2 = X^2 + Y^2$$

Because the priority list requires that the squaring operation be performed before the addition between factors, the operations can be entered just as they appear. That is, for

$$Z = 4 + j8$$
$$Z^2 = 4^2 + 8^2$$

the following is entered using TI-89 calculator:

$$\boxed{4}\ \boxed{\wedge}\ \boxed{2}\ \boxed{+}\ \boxed{8}\ \boxed{\wedge}\ \boxed{2}\ \boxed{\text{ENTER}} \Rightarrow 80$$

However, if the operation to be performed is $Z = \sqrt{4^2 + 8^2}$, then the addition must be performed before the square root, requiring the use of the brackets:

$$Z = \sqrt{(4^2 + 8^2)}$$

$$\boxed{\text{2ND}}\ \sqrt{\ }\ \boxed{4}\ \boxed{\wedge}\ \boxed{2}\ \boxed{+}$$

$$\boxed{8}\ \boxed{\wedge}\ \boxed{2}\ \boxed{)}\ \boxed{\text{ENTER}} \Rightarrow 8.94$$

Trignometric Functions:

All operations such as

$$\sin^{-1}\frac{Y}{Z}, \quad \cos^{-1}\frac{X}{Z}, \quad \text{or} \quad \tan^{-1}\frac{Y}{Z}$$

must be entered as

$$\sin^{-1}\frac{Y}{Z}, \quad \cos^{-1}\frac{X}{Z}, \quad \text{or} \quad \tan^{-1}\frac{Y}{Z}$$

to be sure the division is performed before the inverse operation.

For

$$\theta = \tan^{-1}\frac{3}{8} = \tan^{-1}\left(\frac{3}{8}\right)$$

enter the following for the TI-89 calculator:

$$\boxed{\text{2ND}}\ \boxed{\text{MATH}}\ \boxed{\downarrow}\ \text{A} \downarrow \text{Trig}\ \boxed{\rightarrow}\ \boxed{\downarrow}\ 9 \downarrow \tan^{-1}\ (\ \boxed{\text{ENTER}}$$

$$\boxed{3}\ \boxed{\div}\ \boxed{8}\ \boxed{)}\ \boxed{\text{ENTER}} \Rightarrow 20.56°$$

For cos 30° the following is entered using the TI-89 calculator:

| 2ND | | MATH | | ↓ | A ↓ Trig | → | | ↓ | 2: cos (| ENTER |

| 3 | | 0 | |) | | ENTER | ⇒ 0.866

Basic Equations:

For division in rectangular form, the operation

$$\frac{4}{3^2 + 4^2}$$

must be entered as

$$\frac{4}{(3^2 + 4^2)}$$

resulting in the following sequence using the TI-89 calculator:

| 4 | | ÷ | | (| | 3 | | ∧ | | 2 | | + | | 4 |

| ∧ | | 2 | |) | | ENTER | ⇒ 0.16

EXERCISES

Perform the following operations:

10. $Z = \sqrt{(100)^2 + (40)^2} = $ _____

11. $Z = \sqrt{(4 \times 10^{-8})^2 + (50 \times 10^{-9})^2} = $ _____

12. $Z = \sqrt{(4E3)^2 + (0.08E5)^2} = $ _____

13. $\theta = \sin^{-1} \dfrac{4}{10} = $ _____

14. $\theta = \cos^{-1} \dfrac{1 \times 10^{-3}}{50 \times 10^{-4}} = $ _____

15. $X = \dfrac{4}{\sqrt{4^2 + 6^2}} = $ _____

Complex Numbers

Most scientific calculators provide a sequence of steps for performing the basic operations required with complex numbers, but few are as direct as the TI-86 and TI-89 calculators. The approach is so direct that it is essentially a duplication of the basic operations performed with whole numbers once the complex numbers have been placed in the proper format. The following is for those students with a TI-89 calculator. The steps required for the TI-86 calculator are provided as Appendix in the Text *Introductory Circuit Analysis*, 11th edition. Because time and space do not permit a discussion of the wide variety of calculators available today, for other calculators you will need to consult the operator's manual provided with the calculator.

Conversions: (TI-89 calculator)

Rectangular to Polar form

The conversion process using the TI-89 calculator requires some scrolling to obtain the desired functions. Be assured, however, that with practice, the sequence of steps can be applied quickly and efficiently.

To convert $3 + j4$ to polar form, the sequence appears below. In all cases, complex numbers are entered with surrounding parentheses as shown below. Note also that the calculator using the operator i rather than j.

(3 + 4 2ND *i*) 2ND MATH ↓ 4: Matrix → ↑

L: Vector ops → ↓ 4: ▶ Polar ENTER ENTER (5.00E0 ∠ 53.1E0)

Polar to Rectangular Form

For the reverse process, the complex number must first be identified by the surrounding parentheses, as shown below in the conversion of 10 to the rectangular form.

(1 0 2ND ∠ 2 0) 2ND MATH ↓ 4: Matrix → ↑

L: Vector ops → ↓ 5: ▶ Rect ENTER ENTER (9.40E0 +3.42E0i)

When introducing negative numbers in the complex number format, always remember to take the negative sign from the numeric keybord and not from the list of mathematical operations. In addition, for all operations with complex numbers, be sure to recognize the priority list defined for whole numbers as introduced earlier in the experiment. Finally, in any sequence of operations, always be sure each opening parenthesis has a matching closing parenthesis.

Addition, Subtraction, Multiplication, and Division

For operations such as addition, subtraction, multiplication, and division, simply work with each complex number in the form provided (rectangular or polar) and operate on each in the same manner as whole numbers. When using the TI-89 calculator, the solution will always appear in rectangular form unless the polar form is defined.

As a final example, find the solution to the following mathematical expression and generate an answer in polar form.

$$\frac{5\angle 80°}{(4 + j4) + (8\angle -10°)}$$

The resulting T1-89 sequence is the following:

(5 2ND ∠ 8 0) ÷ ((4

+ 4 2ND *i*) + (8 2ND ∠ (−) 1

0))) 2ND MATH ↓ 4: Matrix → ↑

L: Vector ops → ↓ **4: ▶ Polar** **ENTER** **ENTER** (4.11E−3 ∠67.6E0)

EXERCISES

Perform the following operations:

16. $6\angle-120°$ to rectangular form _____

17. $-5.8 + j9.6$ to polar form _____

18. $6 \times 10^{-4} + j50 \times 10^{-5}$ to polar form _____

19. $12 \times 10^{-3}\angle102°$ to rectangular form _____

20. $(6.8 \text{ mA}\angle-30°)(2.2 \text{ k}\Omega\angle0°) =$ _____ (polar form)

21. $\dfrac{(6.8 \text{ k}\Omega \angle0°)(120 \text{ V} \angle0°)}{6.8 \text{ k}\Omega - j3.6 \text{ k}\Omega} =$ _____ (rectangular form)

22. $5\angle30° + (6 - j6) - 2.2\angle-90° =$ _____ (polar form)

23. $\dfrac{6 + j8}{(2\angle80°)^2 - (6 - j8)} =$ _____ (rectangular form)

EXPERIMENT ac **2**

The Oscilloscope

OBJECTIVES

1. Become familiar with the construction, components, and fundamental operation of an oscilloscope.
2. Learn how to set the amplitude and frequency of a function generator.
3. Understand how to use an oscilloscope to measure both dc and ac voltage levels.
4. Understand the impact of the AC/DC/GND switch on the displayed waveform.

EQUIPMENT REQUIRED

Instruments

1—DMM
1—Oscilloscope
1—Audio oscillator or function generator
1—Frequency counter (if available)

Miscellaneous

2—D batteries with holders

EQUIPMENT ISSUED

TABLE 2.0

Item	Manufacturer and Model No.	Laboratory Serial No.
DMM		
Oscilloscope		
Audio oscillator or function generator		
Frequency counter		

RÉSUMÉ OF THEORY

Oscilloscope

The oscilloscope is an instrument that will display the variation of a voltage with time on a flat screen monitor. The result is that the vertical axis is scaled off in volts while the horizontal axis is in units of time. The number of vertical and horizontal divisions on the screen is not fixed, but the majority have eight vertical divisions and 10 horizontal divisions.

The basic components of an oscilloscope appear in Fig. 2.1. The signal of interest is applied to the vertical input. Depending on its strength, it may be reduced in level (attenuated) or increased in level (amplified). The horizontal input permits applying another signal of any

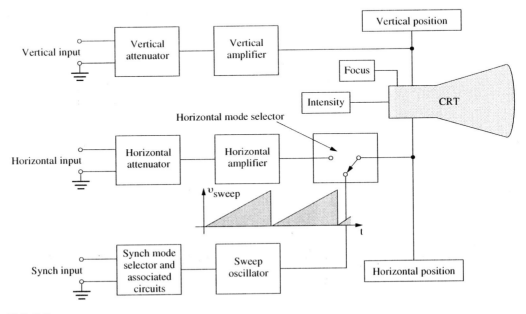

FIG. 2.1

kind to interact with the vertical input to produce a waveform that can often be quite informative. However, if you simply want to view the signal applied to the vertical input versus time, then the sync input option is selected. The sweep oscillator will then generate a sawtooth waveform, such as in Fig. 2.1, to move the applied signal across the screen. If the frequency of the applied signal and that of the sawtooth waveform are the same, the waveforms, are said to be "in sync" and the desired signal will sit stable on the screen. If the two frequencies do not match, the waveform will appear to be continually moving horizontally. Fortunately, all scopes have a sweep time control to adjust the frequency of the sawtooth waveform so a steady image can be displayed.

Note in Fig. 2.1 that the voltage of a sawtooth waveform increases linearly (straight line) with time. This is to ensure that the applied signal will appear across the full width of the screen in an undistorted manner. For the case of an applied sinusoidal voltage, the voltage between the two input terminals will increase and decrease in an oscillatory manner. In the absence of the sawtooth waveform at the horizontal input, the waveform on the screen would simply be a vertical line with a high intensity spot moving up and down on the screen with the same frequency as the applied signal. Applying the sawtooth voltage to the horizontal input will move the waveform across the screen so the full sinusoidal pattern can be displayed.

Most of the controls for the proper operation of an oscilloscope are mounted on the front panel of the instrument. Fig. 2.2 indicates the approximate locations of the controls found on most general-purpose oscilloscopes. The locations of the controls shown vary according to manufacturer. Table 2.1 describes the function of each.

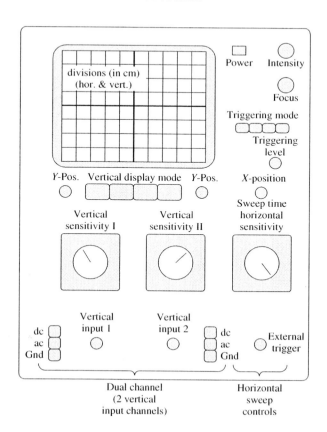

FIG. 2.2

TABLE 2.1

Control	Function
Power	Turns on the main power.
Intensity	Controls the intensity of the pattern on the screen.
Focus	Focuses the electron beam so that the pattern will be clearly defined.
Triggering mode	Determines the type of triggering for the horizontal sweeping pattern.
Triggering level	Determines the level at which triggering should occur.
Vertical display mode	Determines whether one or two signals will be displayed at the same time and which technique will be used to display the signals.
Y-position	Controls the vertical location of the pattern.
X-position	Controls the horizontal location of the pattern.
Vertical sensitivity	Determines the volts/cm for the vertical axis of display.
Sweep time horizontal sensitivity	Determines the time/cm for the horizontal axis of display.
DC/AC/GND switch	Determines whether dc levels will be displayed on the screen and permits determining the GND (0-V input level) of the display.

Basic Measurements

(a) Voltage

dc Levels:

To use the scope to measure dc levels, first place the DC/AC/GND switch in the GND position to establish the ground (0-V) level on the screen.

Then switch the DC/AC/GND switch to the dc position to measure the dc level. In the ac mode, a capacitor blocks the dc from the screen.

Next, place the scope leads across the unknown dc level and use the following equation to determine the dc level:

$$\text{dc Level (V)} = \text{Deflection (div.)} \times \text{Vertical Sensitivity (V/div.)} \qquad \textbf{(2.1)}$$

ac Levels:

After re-establishing the ground level, place the DC/AC/GND switch in the ac mode and connect the scope leads across the unknown voltage. The peak-to-peak voltage can then be determined from

$$V_{p\text{-}p}\,(\text{V}) = \text{Deflection Peak to Peak (div.)} \times \text{Vertical Sensitivity (V/div.)} \qquad \textbf{(2.2)}$$

(b) Frequency

The oscilloscope can be used to set the frequency of an audio oscillator or function generator using the *horizontal sensitivity* in the following manner.

Determine the period of the desired waveform and then calculate the number of divisions required to display the waveform on the horizontal axis using the provided μs/div., ms/div.

or s/div. on the horizontal sensitivity control. Then adjust the audio oscillator or function generator to provide the proper horizontal deflection for the desired frequency.

Of course, the reverse of the above procedure will determine the frequency of an unknown signal.

Audio Oscillator and Function Generator

The *audio oscillator* is designed to provide a sinusoidal waveform in the frequency range *audible* by the human ear. A *function generator* typically expands on the capabilities of the audio oscillator by providing a square wave and triangular waveform with an increased frequency range. Either instrument is suitable for this experiment, since we will be dealing only with sinusoidal waveforms in the audio range.

Most oscillators and generators require that the magnitude of the output signal be set by an oscilloscope or DMM. That is, the amplitude dial of the oscillator is not graduated and the peak or peak-to-peak value is set by connecting the output of the oscillator to a scope or meter and adjusting the amplitude dial until the desired voltage is obtained.

PROCEDURE

Part 1   Introduction

(a) Your instructor will introduce the basic operation of the oscilloscope and audio oscillator or function generator.

(b) Turn on the oscilloscope and establish a horizontal line centered on the face of the screen. There are no connections to the vertical input sections of the scope for this part.

(c) Adjust the controls listed in Table 2.2 and comment on the effects.

TABLE 2.2

Control	Observed Effect
Focus	
Intensity	
Y-position	
X-position	

Part 2   dc Voltage Measurements

(a) Set the DC/AC/GND switch to the GND position and adjust the Y-position control until the 0-V reference is a line centered vertically on the screen.

(b) Once the 0-V level is established, move the DC/AC/GND switch to the dc position and set the vertical sensitivity to 1 V/div. and connect one channel of the scope across the 1.5-V battery as shown in Fig. 2.3.

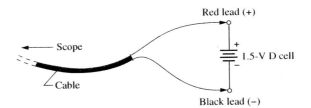

FIG. 2.3

Record the vertical shift below.

Vertical Shift = _____ divisions

Determine the dc voltage that established the shift by multiplying by the vertical sensitivity. That is,

dc Voltage = (Vertical Shift)(Vertical Sensitivity)

= (_____) (_____)

= _____ V

Change the sensitivity to 0.5 V/div. and note the effect on the vertical shift. Recalculate the dc voltage with this new shift.

dc Voltage = (Vertical Shift)(Vertical Sensitivity)

= (_____) (_____)

= _____ V

How do the two measurements compare?

Which is more accurate? Why?

(c) Disconnect the 1.5-V battery and re-establish the 0-V reference line. Then connect the vertical input section of the scope as shown in Fig. 2.4 with the vertical sensitivity set at 1 V/div.

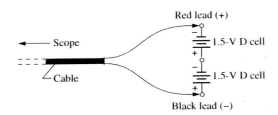

FIG. 2.4

What was the direction of the shift? Why?

Based on the above, can a scope determine the polarity of a voltage? How?

Calculate the magnitude (no sign) of the measured voltage as follows:

dc Voltage = (Vertical Shift)(Vertical Sensitivity)

= (_____) (_____)

= _____ V

Measure the total voltage across the two series batteries with the DMM and compare with the level determined using the oscilloscope.

dc Voltage (DMM) = _____ V

Part 3   Sinusoidal Waveforms—Magnitude

In this part of the experiment, we will learn how to set the magnitude of a sinusoidal signal using an oscilloscope or DMM (or VOM). The frequency will remain fixed at 500 Hz.

Oscilloscope:

(**a**) Connect the output of the oscillator or generator directly to one channel of the scope as shown in Fig. 2.5. If it is available, hook up the frequency counter.

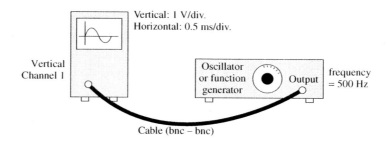

FIG. 2.5

(**b**) Set the output frequency of the oscillator or generator to 500 Hz using the dial and appropriate multiplier. Turn the amplitude knob all the way to the left for minimum output.

(c) Set the vertical sensitivity of the scope to 1 V/div. and the horizontal sensitivity to 0.5 ms/div. and turn on both the scope and the oscillator or generator.

(d) Set the DC/AC/GND switch to the GND position to establish the 0-V reference level (also the vertical center of a sinusoidal waveform) and then return the switch to the ac position.

(e) Now adjust the amplitude control of the oscillator or generator until the signal has a 6-V peak-to-peak swing. The resulting waveform has the following mathematical formulation:

$$v = V_m \sin 2\pi ft = 3 \sin 2\pi 500t$$

(f) Switch to the dc position and comment below on any change in location or appearance of the waveform.

(g) Make the necessary adjustments to display the following waveforms on the screen. Sketch both patterns in Figs. 2.6 and 2.7, showing the number of divisions (in centimeters) for the vertical and horizontal distances, and the vertical and horizontal sensitivities. Use a sharp pencil for the sketch. Be neat and accurate!

1. $v = 0.2 \sin 2\pi 500t$

   Vertical Sensitivity = _____

   Horizontal Sensitivity = _____

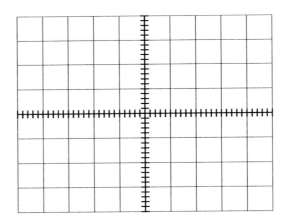

FIG. 2.6

   2. $v = 8 \sin 2\pi 500t$

   Vertical Sensitivity = _____

   Horizontal Sensitivity = _____

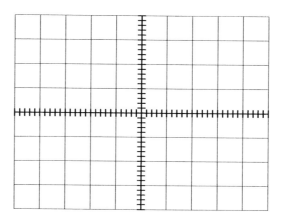

FIG. 2.7

DMM:

   (h) The sinusoidal signal $3 \sin 2\pi 500t$ has an effective value determined by

$$V_{\text{eff}} = 0.707V_m = 0.707(3 \text{ V}) = 2.121 \text{ V}$$

   Connect the DMM directly across the oscillator in the ac rms mode and adjust the oscillator output until $V_{\text{eff}} = 2.121$ V. Then connect the output of the oscillator directly to the scope and note the total peak-to-peak swing.

   Is the waveform the same as that obtained in part 3(e)?

   (i) Use the DMM to set the following sinusoidal output from the oscillator:

$v = 0.5 \sin 2\pi 500t$

$V_{\text{eff}} = $ _____ V

   Set V_{eff} with the DMM by adjusting the output of the oscillator, and place the signal on the screen.

   Calculate the peak-to-peak voltage as follows:

$V_{p\text{-}p}$ = (vertical distance peak to peak)(vertical sensitivity)

   = (_____) (_____)

   = _____ V

   How does the above compare with the desired 1-V peak-to-peak voltage?

Part 4   Sinusoidal Waveforms—Frequency

This section will demonstrate how the oscilloscope can be used to set the frequency output of an oscillator or generator. In other words, the scope can be used to make fine adjustments on the frequency set by the dials of the oscillator or generator.

For a signal such as $2 \sin 2\pi 500t$, the frequency is 500 Hz and the period is 1/500 Hz = 2 ms. With a horizontal sensitivity of 0.5 ms/div., the waveform should appear in exactly four horizontal divisions. If it does not, the fine-adjust control on the frequency of the oscillator or gen-erator can be adjusted until it is exactly 4 divisions. The scope has then set the output frequency of the oscillator.

Make the necessary adjustments to place the following waveforms on the scope. Sketch the waveforms on the scope patterns in Figs. 2.8 and 2.9, indicating the number of vertical and horizontal deflections and the sensitivity of each. Use a frequency counter if one is available.

    1. $v = 0.4 \sin 62{,}832t$

        $f =$ _____ Hz, $T =$ _____ s

        Vertical Deflection (peak value) = _____ divisions

        Vertical Sensitivity = _____

        Horizontal Deflection (for one period of waveform) = _____ divisions

        Horizontal Sensitivity = _____

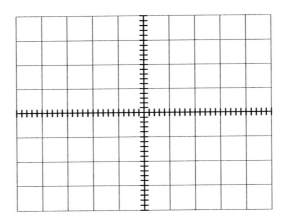

FIG. 2.8

2. $v = 5 \sin 377t$

$f =$ _____ Hz, $T =$ _____ s

Vertical Deflection (peak value) = _____ divisions

Vertical Sensitivity = _____

Horizontal Deflection (for one period of waveform) = _____ divisions

Horizontal Sensitivity = _____

FIG. 2.9

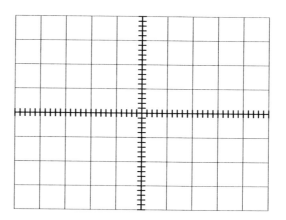

Part 5 Sinusoidal Waveforms on a dc Level

(a) Set the oscillator or generator to an output of 1 sin $2\pi500t$ using a vertical sensitivity of 1 V/div. on the scope with a horizontal sensitivity of 0.5 ms/div.

(b) Measure the dc voltage of one of the D cells and insert in Fig. 2.10.

$E =$ _____ V

(c) Construct the series combination of supplies as shown in Fig. 2.10 and connect the scope as indicated.

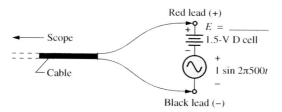

Red lead (+)

Scope

$E =$ _____

1.5-V D cell

Cable

1 sin $2\pi500t$

Black lead (−)

FIG. 2.10

(d) The input signal now has a dc level equal to the dc voltage of the D cell.

Set the DC/AC/GND switch to the GND position and adjust the zero line to the center of the screen.

(e) Switch to the AC mode and sketch the waveform on Fig. 2.11.

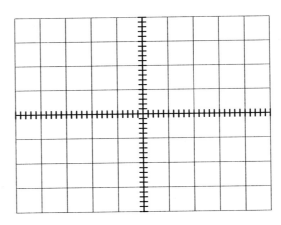

FIG. 2.11

(f) Now switch to the DC mode and sketch the waveform on the same scope pattern as part 5(e).

(g) What was the effect of switching from the AC to the DC mode?

Did the shape of the sinusoidal pattern change at all?

How does the vertical shift compare to the dc level of the battery?

(h) Switch to the GND mode and describe what happened to the waveform. In general, what is the effect of switching to the GND position, no matter where the leads of the scope are connected?

EXERCISE

1. Write the sinusoidal expression for the waveform appearing in Fig. 2.12.

   $v =$ _____

   Vertical
   sensitivity = 4 V/div.

   Horizontal
   sensitivity = 5 μs/div.

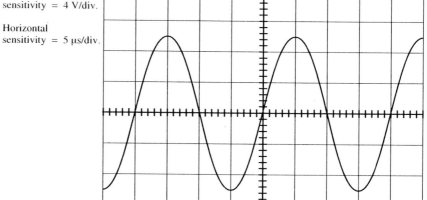

FIG. 2.12

2. Sketch the waveform defined by

   $v = -1.5 + 2.5 \sin 2\pi(20 \times 10^3)t$

   on the scope pattern of Fig. 2.13. Include the vertical and horizontal sensitivities.

   Vertical Sensitivity = _____
   Horizontal Sensitivity = _____

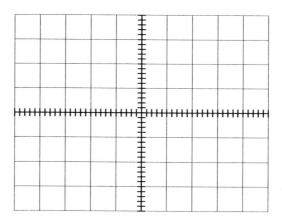

FIG. 2.13

R-L-C Components

OBJECTIVES

1. Develop skills with the oscilloscope as a voltage-measuring instrument.
2. Learn how to measure the impedance of an element using a current-sensing resistor.
3. Compare the measured and nameplate values of a resistor, inductor, and capacitor.

EQUIPMENT REQUIRED

Resistors

1—100-Ω, 1.2-kΩ, 3.3-kΩ (1/4-W)

Inductors

2—10-mH

Capacitors

1—0.47-μF, 1-μF

Instruments

1—DMM

1—Oscilloscope

1—Audio oscillator (or signal generator)

1—Frequency counter (if available)

EQUIPMENT ISSUED

TABLE 3.0

Item	Manufacturer and Model No.	Laboratory Serial No.
DMM		
Oscilloscope		
Audio oscillator (or signal generator)		
Frequency counter		

RÉSUMÉ OF THEORY

The reactance of an inductor or capacitor is a function of the applied frequency as defined by Eqs. (3.1) and (3.2).

$$X_L = 2\pi f L \tag{3.1}$$

$$X_C = \frac{1}{2\pi f C} \tag{3.2}$$

For elements in series, the total impedance is the sum of the individual impedances:

$$\mathbf{Z}_T = \mathbf{Z}_1 + \mathbf{Z}_2 + \mathbf{Z}_3 + \cdots + \mathbf{Z}_N \tag{3.3}$$

For resistors in series,

$$R_T = R_1 + R_2 + R_3 + \cdots + R_N \tag{3.4}$$

which is independent of frequency.

For inductors in series,

$$X_{L_T} = 2\pi f L_T$$

with

$$L_T = L_1 + L_2 + L_3 + \cdots + L_N \tag{3.5}$$

Note that the individual and total inductive reactances are directly proportional to frequency.

For capacitors in series,

$$X_{C_T} = \frac{1}{2\pi f C_T}$$

with

$$\frac{1}{C_T} = \frac{1}{C_1} + \frac{1}{C_2} + \frac{1}{C_3} + \cdots + \frac{1}{C_N} \tag{3.6}$$

Note that the individual and total capacitive reactances are inversely proportional to frequency.

For the special case of only two capacitors in series:

$$X_{C_T} = \frac{1}{2\pi f C_T}$$

where

$$C_T = \frac{C_1 C_2}{C_1 + C_2}$$

For resistors in parallel,

$$\boxed{\frac{1}{R_T} = \frac{1}{R_1} + \frac{1}{R_2} + \frac{1}{R_3} + \cdots + \frac{1}{R_N}} \tag{3.7}$$

which is independent of frequency.

For inductors in parallel,

$$X_{L_T} = 2\pi f L_T$$

with

$$\boxed{\frac{1}{L_T} = \frac{1}{L_1} + \frac{1}{L_2} + \frac{1}{L_3} + \cdots + \frac{1}{L_N}} \tag{3.8}$$

For the special case of only two inductors in parallel:

$$X_{L_T} = 2\pi f L_T$$

where

$$L_T = \frac{L_1 L_2}{L_1 + L_2}$$

For capacitors in parallel,

$$X_{C_T} = \frac{1}{2\pi f C_T}$$

where

$$\boxed{C_T = C_1 + C_2 + C_3 + \cdots + C_N} \tag{3.9}$$

For the special case of only two capacitors in parallel:

$$X_{C_T} = \frac{1}{2\pi f C_T}$$

where

$$C_T = C_1 + C_2$$

PROCEDURE

Part 1   Resistance

(a) Construct the circuit of Fig. 3.1. Insert the actual values of the resistors as determined by the ohmmeter section of your multimeter. Hook up the frequency counter if available.

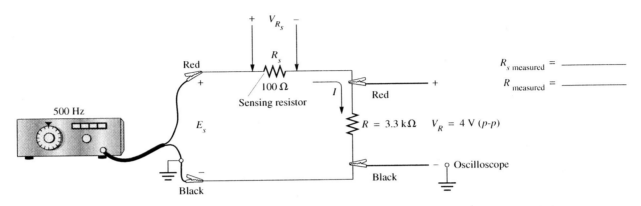

FIG. 3.1

Caution: **Always ensure that the ground of the oscilloscope is connected to the ground of the oscillator. Otherwise a hazardous situation may result.**

(b) Set the voltage across R to 4 V (p-p) by adjusting the source voltage E_S and observing V_R with the oscilloscope. Measure the rms voltage across the sensing resistor (100 Ω) with the DMM, then calculate the peak value of V_{R_S} and record both results in Table 3.1.

Calculation:

TABLE 3.1

V_{R_S}(DMM)	$V_{R_{S(peak)}}$	$V_{R_{S(p\text{-}p)}}$	$I_{p\text{-}p}$	R

Calculate the peak-to-peak value of V_{R_S} and record in Table 3.1.

Calculation:

(c) Calculate the peak-to-peak value of the current I from

$$I_{p\text{-}p} = \frac{V_{R_{s(p\text{-}p)}}}{R_s}$$

and record in Table 3.1.

Calculation:

(d) Determine the resistance of the resistor R from

$$R = \frac{4\,V_{(p\text{-}p)}}{I_{p\text{-}p}}$$

and record in Table 3.1.

Calculation:

(e) Compare the value obtained in part 1(d) with the measured value inserted in Fig. 3.1.

(f) Connect a 1.2-kΩ and a 3.3-kΩ resistor in series as shown in Fig. 3.2. Record the measured values of each resistor in the space provided.

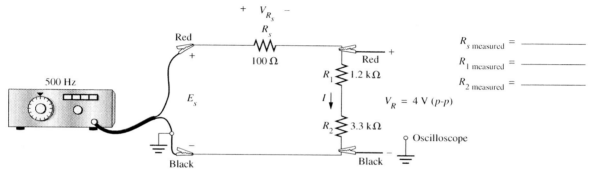

FIG. 3.2

Calculate the total resistance of the series combination of the 1.2-kΩ and 3.3-kΩ resistors using the measured values and record in Table 3.2.

TABLE 3.2

R_T (calc.)	V_{R_s} (DMM)	$V_{R_{s(p\text{-}p)}}$	$I_{p\text{-}p}$	R_T

(g) Set the voltage across the series combination to 4 V (p-p) and then measure the voltage V_{R_s} with the DMM and record in Table 3.2.

(h) Calculate the peak-to-peak value of V_{R_s} and record in Table 3.2.

Calculation:

(i) Determine the peak-to-peak value of the current I from

$$I_{p\text{-}p} = \frac{V_{R_{s(p\text{-}p)}}}{R_s}$$

and record in Table 3.2.

Calculation:

(j) Calculate the total resistance R_T of the series combination of R_1 and R_2 from

$$R_T = \frac{V_{R_{T(p\text{-}p)}}}{I_{p\text{-}p}} = \frac{4 \text{ V}}{I_{p\text{-}p}}$$

and record in the last column of Table 3.2.

Calculation:

(k) Compare the results of part 1(j) with the calculated value of part 1(f).

Part 2 Capacitive Reactance

(a) Construct the circuit of Fig. 3.3. Insert the measured value of R_s.

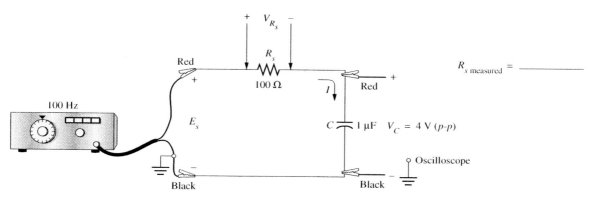

FIG. 3.3

(b) Set the voltage V_C to 4 V (*p-p*) and measure the rms value of the voltage V_{R_s} with the DMM. Record in the top row of Table 3.3.

(c) Using the technique described in part 1, calculate the peak-to-peak value of the current I and record in Table 3.3.

Calculation:

TABLE 3.3

	V_{R_s} **(DMM)**	$I_{p\text{-}p}$	X_C **(meas.)**	X_C **(calc.)**	C
Parts (b)–(f)					
Parts (g)–(j)					

(d) Calculate the reactance X_C from the peak-to-peak values of V_C and I and record in Table 3.3.

$$X_C = \frac{V_{C(p\text{-}p)}}{I_{p\text{-}p}}$$

Calculation:

(e) Using the nameplate value of the capacitance (1 μF), calculate the reactance of the capacitor at $f = 100$ Hz using Eq. (3.2) and record in Table 3.3. Compare the results with that of part 2(d).

Calculation:

Comparison:

(f) Using the measured value of X_C in part 2(d), determine the capacitance level using Eq. (3.2) if $f = 100$ Hz and record in Table 3.3. Compare with the nameplate value (1 μF) of the capacitance.

Calculation:

(g) Connect a 0.47-μF in parallel with the capacitor of Fig. 3.3 and set V_C to 4 V (p-p) again. Measure the rms value of V_{R_S} using a DMM and record in the bottom row of Table 3.3.

Calculate the peak-to-peak value of V_{R_S} and determine the peak-to-peak value of I from

$$I_{p\text{-}p} = \frac{V_{R_{s(p\text{-}p)}}}{R_s}$$

and record in Table 3.3.

Calculation:

(**h**) Calculate the total reactance of the parallel capacitors from

$$X_{C_T} = \frac{V_{C(p\text{-}p)}}{I_{p\text{-}p}} = \frac{4 \text{ V}}{I_{p\text{-}p}}$$

and record in Table 3.3.

Calculation:

(**i**) Using the total capacitance as determined from the nameplate values and Eq. (3.2), calculate the total reactance of the parallel capacitor at a frequency of $f = 100$ Hz and record in Table 3.3.

Calculation:

Compare the result with the measured value of part 2(h).

(**j**) Using the measured value of X_{C_T} in part 2(h), calculate the equivalence total capacitance at $f = 100$ Hz and record in Table 3.3. Compare with the nameplate level.

Calculation:

Comparison:

Part 3   Inductive Reactance

(a) Construct the network of Fig. 3.4. Insert the measured value for R_s. The internal resistance of the coil will be ignored in this experiment since it is considerably less than X_L at the applied frequency.

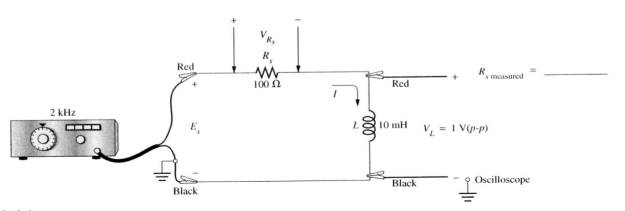

FIG. 3.4

(b) Set the voltage V_L to 1 V (p-p) and measure the rms value of the voltage V_{R_s} with a DMM. Record in the top row of Table 3.4.

(c) Calculate $V_{R_{s(p-p)}}$ and record in Table 3.4.

Calculation:

TABLE 3.4

	V_{R_S} (DMM)	$V_{R_{S(p\text{-}p)}}$	$I_{p\text{-}p}$	X_L (meas.)	X_L (calc.)	L
parts (b)–(i)						
parts (j)–(n)						

    (d) Calculate $I_{p\text{-}p}$ and record in Table 3.4.

Calculation:

    (e) Calculate the reactance X_L from

$$X_L = \frac{V_{L(p\text{-}p)}}{I_{p\text{-}p}} = \frac{1\text{ V}}{I_{p\text{-}p}}$$

and record in Table 3.4.

Calculation:

    (f) Using the nameplate value of inductance L and Eq. (3.1), calculate the inductive reactance at $f = 2$ kHz and record in Table 3.4.

Calculation:

    (g) Compare the results of part 3(f) with the measured level of part 3(e).

    (h) Calculate the inductance level necessary to establish an inductive reactance of the magnitude measured in part 3(e) using Eq. (3.1) and record in the last column of Table 3.4.

Calculation:

(i) Compare the result of part 3(h) with the nameplate inductance level.

(j) Connect a second 10-mH coil in series with the coil of Fig. 3.4 and establish 1 V (p-p) across the two series coils. Then measure the rms value of V_{R_s} and calculate $V_{R_{s(p-p)}}$ and record the results in the bottom row of Table 3.4.

Calculation:

(k) Calculate I_{p-p} and record in Table 3.4.

Calculation:

(l) Determine X_{L_T} from

$$X_{L_T} = \frac{V_{L(p-p)}}{I_{p-p}} = \frac{1\ \text{V}}{I_{p-p}}$$

and record in Table 3.4.

Calculation:

(m) Using the nameplate values of the inductances and Eq. (3.1), calculate the total reactance at $f = 2$ kHz and compare with the result of part 3(l).

Calculation:

Comparison:

    (n) Using the measured value of X_{L_T} in part 3(l) and Eq. (3.1), calculate the required inductance level at $f = 2$ kHz and record in Table 3.4. Compare with the total nameplate inductance level.

Calculation:

Comparison:

EXERCISES

    1. Given the information provided in Fig. 3.5, determine the level of R.

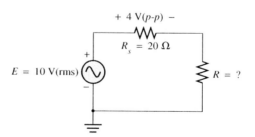

FIG. 3.5

$R =$ _____

2. Given the information provided in Fig. 3.6, determine the level of inductance L.

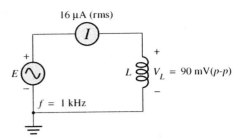

16 μA (rms)

E

$f = 1$ kHz

L

$V_L = 90$ mV(p-p)

FIG. 3.6

$R = \underline{\qquad\qquad}$

3. Given the information provided in Fig. 3.7, determine the level of capacitance C.

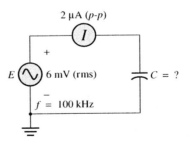

2 μA (p-p)

E

6 mV (rms)

$f = 100$ kHz

$C = ?$

FIG. 3.7

$C = \underline{\qquad\qquad}$

Name:_____

Date:_____

Course and Section:_____

Instructor:_____

Frequency Response of *R, L,* and *C* Components

OBJECTIVES

1. Verify that the resistance of a resistor is independent of frequency for frequencies in the audio range.
2. Note that the reactance of an inductor increases linearly with increase in frequency.
3. Confirm that the reactance of a capacitor decreases nonlinearly with increase in frequency.

EQUIPMENT REQUIRED

Resistors

1—100-Ω, 1-kΩ (1/4-W)

Inductors

1—10-mH

Capacitors

1—0.1-μF, 0.47-μF

Instruments

1—DMM

1—Oscilloscope

1—Audio oscillator (or function generator)

1—Frequency counter (if available)

EQUIPMENT ISSUED

TABLE 4.0

Item	Manufacturer and Model No.	Laboratory Serial No.
DMM		
Oscilloscope		
Audio oscillator (or function generator)		
Frequency counter		

RÉSUMÉ OF THEORY

The resistance of a carbon resistor is unaffected by frequency, except for extremely high frequencies. This rule is also true for the total resistance of resistors in series or parallel.

The reactance of an inductor is linearly dependent on the frequency applied. That is, if we double the frequency, we double the reactance, as determined by $X_L = 2\pi fL$. For very low frequencies, the reactance is correspondingly very small, whereas for increasing frequencies, the reactance will increase to a very large value. For dc conditions, we find that $X_L = 2\pi(0)L$ is $0\ \Omega$, corresponding with the short-circuit representation we used in our analysis of dc circuits. For very high frequencies, X_L is so high that we can often use an open-circuit approximation.

At low frequencies, the reactance of a coil is quite low, whereas for a capacitor the reactance is quite high at low frequencies, often permitting the use of an open-circuit equivalent. At higher frequencies, the reactance of a coil increases rapidly in a linear fashion, but the reactance of a capacitor decreases in a nonlinear manner. In fact, it drops off more rapidly than the reactance of a coil increases. At very high frequencies, the capacitor can be approximated by a short-circuit equivalency.

PROCEDURE

Part 1   Resistors

Construct the circuit of Fig. 4.1. Insert the measured value of R. Hook up the freqency counter if available.

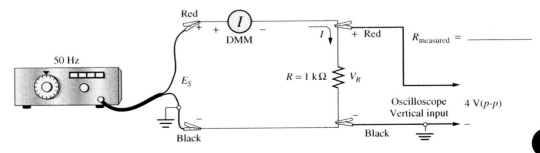

FIG. 4.1

In this part of the experiment, the voltage across the resistor will be held constant at 4 V (p-p) while only the frequency is varied. If the resistance is frequency independent, the current through the circuit should not change as a function of frequency. Therefore, by keeping the voltage V_R constant and changing the frequency while monitoring the current I, we can verify if indeed resistance is frequency independent.

Set the voltage V_R across the 1-kΩ resistor to 4 V (p-p) using the oscilloscope. Note that the first frequency of Table 4.2 is 50 Hz. In addition, note that it is the voltage across the resistor that is set to 4 V (p-p), not the supply voltage. For each frequency of Table 4.2, be sure V_R is maintained at 4 V (p-p) as the rms level of the current is measured using the DMM.

TABLE 4.1

Frequency	$V_{R(p\text{-}p)}$	Calculation: $V_{R(rms)} = 0.707\left(\dfrac{V_{R(p\text{-}p)}}{2}\right)$	Measurement: I_{rms}	Calculation: $R = \dfrac{V_{R(rms)}}{I_{rms}}$
50 Hz	4 V	1.414 V		
100 Hz	4 V	1.414 V		
200 Hz	4 V	1.414 V		
500 Hz	4 V	1.414 V		
1000 Hz	4 V	1.414 V		

Calculate the level of R at each frequency using Ohm's law and complete Table 4.1. Use the following space for your calculations. Based on the results of Table 4.1, is the resistance of the resistor independent of frequency for the tested range?

Part 2   Inductors

Construct the circuit of Fig. 4.2. The dc resistance of the coil (R_l) will be ignored for this experiment, because $X_L >> R_l$. Insert the measured value of R_s and hook up the frequency counter if available.

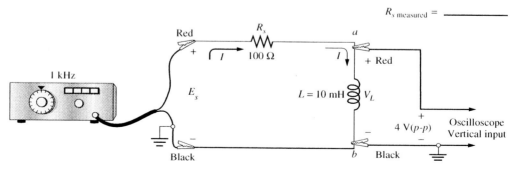

FIG. 4.2

In this part, the resistor of part 1 is replaced by the inductor. Here again, the voltage across the inductor will be kept constant while we vary the frequency of that voltage and monitor the current in the circuit.

Set the frequency of the function generator to 1 kHz and adjust E_s until the voltage across the coil (V_L) is 4 V (p-p). Then turn off the supply without touching its controls and interchange the positions of the sensing resistor R_s and the inductor. The purpose of this maneuver is to ensure a common ground between the oscilloscope and the supply. Turn on the supply and measure the peak-to-peak voltage V_{R_s} across the sensing resistor. Use Ohm's law to determine the peak-to-peak value of the current through the series circuit and insert in Table 4.2. Repeat the above for each frequency appearing in Table 4.2.

TABLE 4.2

Frequency	$V_{L(p\text{-}p)}$	$V_{Rs(p\text{-}p)}$	$I_{p\text{-}p} = \dfrac{V_{R_{s(p\text{-}p)}}}{R_s(\text{meas.})}$	$X_L(\text{measured}) = \dfrac{V_{L(p\text{-}p)}}{I_{p\text{-}p}}$	$X_L \text{ (calculated)} = 2\pi fL$
1 kHz	4 V				
3 kHz	4 V				
5 kHz	4 V				
7 kHz	4 V				
10 kHz	4 V				

The DMM was not used to measure the current in this part of the experiment because many commercial units are limited to frequencies of 1 kHz or less.

(a) Calculate the reactance X_L (magnitude only) at each frequency and insert the values in Table 4.3 under the heading "X_L (measured)."

(b) Calculate the reactance at each frequency of Table 4.2 using the nameplate value of inductance (10 mH), and complete the table.

(c) How do the measured and calculated values of X_L compare?

(d) Plot the measured value of X_L versus frequency on Graph 4.1. Label the curve and plot the points accurately. Include the plot point of $f = 0$ Hz and $X_L = 0\ \Omega$ as determined by $X_L = 2\pi fL = 2\pi(0\ \text{Hz})L = 0\ \Omega$.

(e) Is the resulting plot a straight line? Should it be? Why?

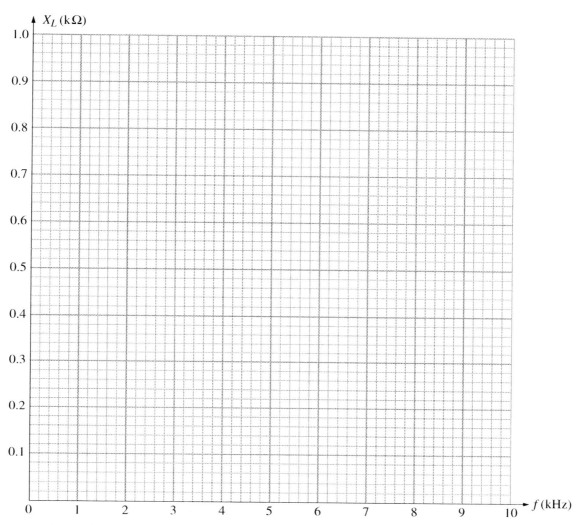

GRAPH 4.1

(f) Determine the inductance at 1.5 kHz using the plot of part 2(d). That is, determine X_L from the graph at $f = 1.5$ kHz, calculate L from $L = X_L/2\pi f$ and insert the results in Table 4.3.

Calculation:

TABLE 4.3

X_L	L (calc.)	L (nameplate)

Record the nameplate value of the coil in Table 4.3. Compare the nameplate value to the calculated value.

Part 3   Capacitors

(a) Construct the circuit of Fig. 4.3.

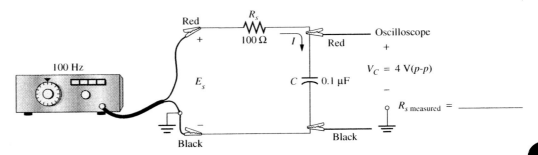

FIG. 4.3

The 100-Ω sensing resistor will be used to "sense" the current level in the network. With the generator set at 100 Hz, adjust the output of supply until a 4-V peak-to-peak signal is obtained across the capacitor. Then repeat the procedure as outlined in part 2(a) to determine $I_{p\text{-}p}$ for each frequency appearing in Table 4.4.

(b) Calculate X_C from the measured values at each frequency and insert in the X_C (measured) column of Table 4.4, since both V_C and I are measured values.

TABLE 4.4

Frequency	$V_{C(p\text{-}p)}$	$V_{R_{s(p\text{-}p)}}$	$I_{p\text{-}p}$	X_C (measured) $= \dfrac{V_{C(p\text{-}p)}}{I_{p\text{-}p}}$	X_C (calculated) $= \dfrac{1}{2\pi fC}$
100 Hz	4 V				
200 Hz	4 V				
300 Hz	4 V				
400 Hz	4 V				
500 Hz	4 V				
800 Hz	4 V				
1000 Hz	4 V				
2000 Hz	4 V				

(c) Calculate X_C using the nameplate capacitance level of 0.1 μF at each frequency and insert in the X_C (calculated) column.

(d) How do the results in the X_C (measured) column compare with those in the X_C (calculated) column?

(e) Plot X_C (measured) versus frequency on Graph 4.2. Extend the curve below the lowest measured frequency level using a calculated or estimated level of X_C. Label the curve and clearly show all the data points.

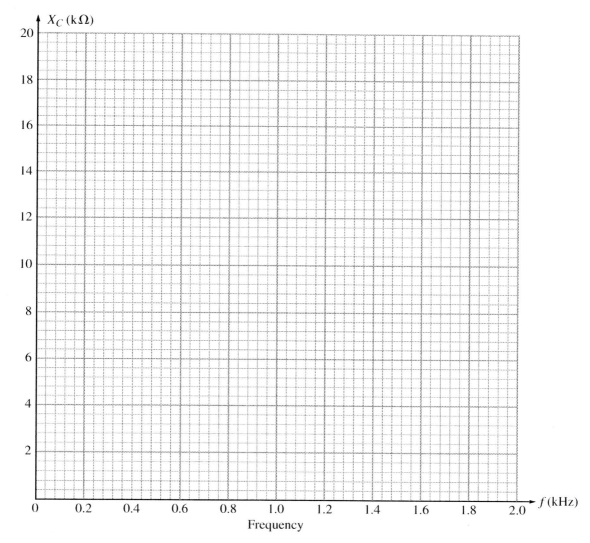

GRAPH 4.2

In what frequency range does the greatest change in X_C occur?

Is the graph linear or nonlinear?

How would you compare the curve of X_C versus frequency to that of X_L versus frequency from part 2?

(f) Determine X_C at a frequency of 650 Hz from the curve just plotted and insert in Table 4.5. At this frequency, determine the equivalent capacitance level using the fact that $C = 1/2\pi f X_C$ as defined by $X_C = 1/2\pi fC$. Insert the result in Table 4.5 along with the nameplate value of the capacitor.

Calculation:

TABLE 4.5

X_C	C (calc.)	C (nameplate)

How does the calculated level compare with the nameplate level?

Part 4 Parallel Capacitors (Increase in Capacitance)

(a) Place a 0.47-µF capacitor in parallel with the 0.1-µF capacitor of Fig. 4.3 to obtain a nameplate capacitance level of 0.57 µF ($C_T = C_1 + C_2$).

(b) Maintaining 4 V (*p-p*) across the parallel capacitors, measure and insert the peak-to-peak value of V_{R_s} for each frequency in Table 4.6 using the technique described in the earlier parts of the experiment.

TABLE 4.6

Frequency	$V_{C(p\text{-}p)}$	$V_{R_{s(p\text{-}p)}}$	$I_{p\text{-}p}$	$X_C \text{ (measured)} = \dfrac{V_{C(p\text{-}p)}}{I_{p\text{-}p}}$	$X_C \text{ (calculated)} = \dfrac{1}{2\pi fC}$
50 Hz	4 V				
100 Hz	4 V				
200 Hz	4 V				
300 Hz	4 V				
400 Hz	4 V				
500 Hz	4 V				
800 Hz	4 V				
1000 Hz	4 V				
2000 Hz	4 V				

(c) For each frequency in Table 4.6, calculate the peak-to-peak value of the current from

$$I_{p\text{-}p} = \frac{V_{R_{s(p\text{-}p)}}}{R_s \text{(measured)}}$$

and insert in the $I_{p\text{-}p}$ column of Table 4.6.

(d) Calculate X_C from the measured values at each frequency and insert in the X_C (measured) column of Table 4.6.

(e) Calculate X_C using the nameplate capacitance level of 0.57 µF at each frequency and insert in Table 4.6 in the X_C (calculated) column.

(f) How do the results in the X_C (measured) column compare with those in the X_C (calculated) column?

(g) Plot X_C (measured) versus frequency on Graph 4.2. How did the change (increase) in capacitance level affect the location and characteristics of the curve?

(h) Determine X_C at a frequency of 300 Hz from the curve just plotted and record in Table 4.7. At this frequency, determine the equivalent capacitance level using $C = 1/2\pi f X_C$ and record in Table 4.7 along with the total nameplate value.

Calculation:

TABLE 4.7

X_C	C (calc.)	C (nameplate)

How does the calculated capacitance level compare with the nameplate level?

EXERCISES

1. In the experiment, the effect of an increase in capacitance on the X_C curve was investigated. Note the effect of an increase in inductance on the X_L curve. Increase the inductance of Fig. 4.2 to 20 mH and plot the curve of X_L versus frequency on Graph 4.1 for a frequency range of 0 Hz to 10 kHz. Be sure to label the curve and clearly indicate the plot points chosen.

   What was the effect on the X_L curve due to an increase in inductance?

2. Determine the level of *C* to establish the voltage levels of Fig. 4.4. Show all calculations and organize your work.

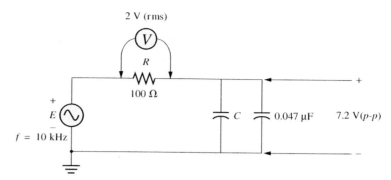

FIG. 4.4

C = _____

3. Determine the level of *L* to establish the voltage levels of Fig. 4.5. Again, show all required calculations and organize your work.

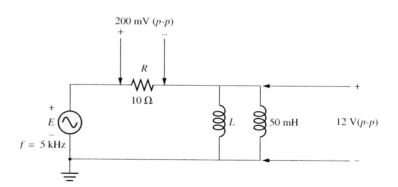

FIG. 4.5

L = _____

4. Using PSpice or Multisim, obtain a plot of X_L versus frequency for the 10 mH coil of Fig. 4.2. Compare the resulting curve to that obtained on Graph 4.1. How does the value of X_L at $f = 1.5$ kHz compare with the values appearing in Table 4.3?

  Attach all appropriate computer printouts.

EXPERIMENT ac **5**

Frequency Response of the Series *R-L* Network

OBJECTIVES

1. Note the effect of frequency on the impedance of a series *R-L* network.
2. Plot the voltages and current of a series *R-L* network versus frequency.
3. Calculate and plot the phase angle of the input impedance versus frequency for a series *R-L* network.

EQUIPMENT REQUIRED

Resistors

1—100-Ω (1/4-W)

Inductors

1—10-mH

Instruments

1—DMM

1—Oscilloscope

1—Audio oscillator (or function generator)

1—Frequency counter (if available)

EQUIPMENT ISSUED

TABLE 5.0

Item	Manufacturer and Model No.	Laboratory Serial No.
DMM		
Oscilloscope		
Audio oscillator (or function generator)		
Frequency counter		

RÉSUMÉ OF THEORY

For the series dc or ac circuit, the voltage drop across a particular element is directly related to its impedance as compared with the other series elements. Since the impedances of the inductor and capacitor will change with frequency, the voltage across both elements will be affected by the applied frequency.

For the series R-L network, the voltage across the coil will increase with frequency, since the inductive reactance increases directly with frequency and the impedance of the resistor is essentially independent of the applied frequency (in the audio range).

Since the voltage and current of the resistor are related by the fixed resistance value, the shapes of their curves versus frequency will have the same characteristics.

Keep in mind that the voltages across the elements in an ac circuit are vectorially related. Otherwise, the voltage readings may appear to be totally incorrect and not satisfy Kirchhoff's voltage law.

The phase angle associated with the input impedance is also sensitive to the applied frequency. At very low frequencies, the inductive reactance will be small compared to the series resistive element and the network will be primarily resistive in nature. The result is a phase angle associated with the input impedance that approaches $0°$ (v and i in phase). At increasing frequencies, X_L will drown out the resistive element and the network will be primarily inductive, resulting in an input phase angle approaching $90°$ (v leads i by $90°$).

Caution: **Be sure that the ground connections of the source and scope do not short out an element of the network, thereby changing its terminal characteristics.**

PROCEDURE

Part 1 V_L, V_R, and I versus Frequency

(a) Construct the network of Fig. 5.1. Insert the measured value of the resistor R on the diagram. For the frequency range of interest, we will ignore the effects of the internal resistance of the coil. That is, we will assume $X_L >> R_l$ and $\mathbf{Z}_L = X_L \angle 90°$.

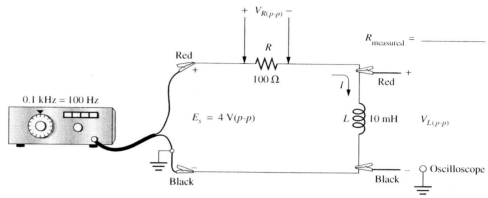

FIG. 5.1

(b) Maintaining 4 V (*p-p*) at the input to the circuit, record the voltage V_L (*p-p*) for the frequencies appearing in Table 5.1. Make sure to check continually that $E = 4$ V (*p-p*) with each frequency change. Do not measure the voltage V_R at this point in the experiment. The common ground of the supply and scope will short out the effect of the inductive element, which may result in damage to the equipment.

For each frequency try to read V_L to the highest degree of accuracy possible. The higher the degree of accuracy, the better the data will verify the theory to be substantiated.

TABLE 5.1

Frequency	$V_{L(p\text{-}p)}$	$V_{R(p\text{-}p)}$	$I_{p\text{-}p}$
0.1 kHz			
1 kHz			
2 kHz			
3 kHz			
4 kHz			
5 kHz			
6 kHz			
7 kHz			
8 kHz			
9 kHz			
10 kHz			

(c) Turn off the supply and interchange the positions of *R* and *L* in Fig. 5.1 and measure $V_{R(p\text{-}p)}$ for the same range of frequencies with *E* maintained at 4 V (*p-p*). Insert the measurements in Table 5.1. **This is a very important step.** Failure to relocate the resistor *R* can result in a grounding situation where the inductive reactance is shorted out!

(d) Calculate $I_{p\text{-}p} = V_{R(p\text{-}p)}/R_{\text{measured}}$ and complete Table 5.1.

(e) Plot the curve of $V_{L(p\text{-}p)}$ versus frequency on Graph 5.1. Label the curve and clearly indicate each plot point.

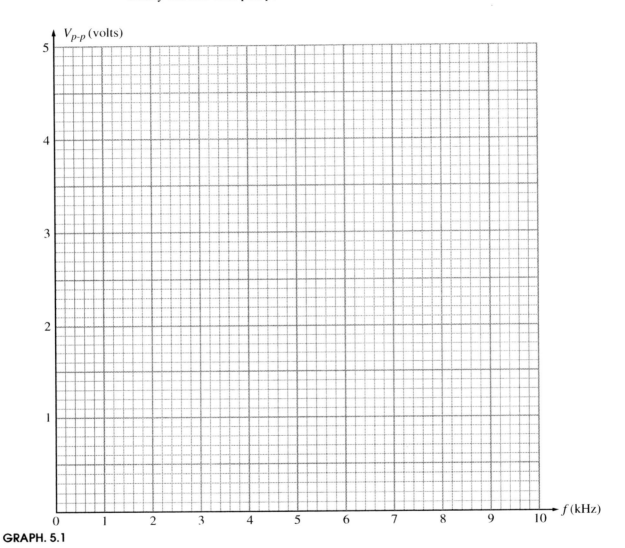

GRAPH. 5.1

(f) Plot the curve of $V_{R(p\text{-}p)}$ versus frequency on Graph 5.1. Again, label the curve and clearly indicate each plot point.

(g) As the frequency increases, describe in a few sentences what happens to the voltage across the coil and resistor. Explain why.

(h) At the point where $V_L = V_R$, does $X_L = R$? Should they be equal? Why? Record the level of voltage and the impedance of each element in Table 5.2.

TABLE 5.2

$V(V_L = V_R)$	X_L	R

(i) Determine $V_{L(p\text{-}p)}$ and $V_{R(p\text{-}p)}$ at some random frequency such as 5.6 kHz from the curves and record in Table 5.3. Determine their sum and enter into Table 5.3.

Calculation:

Are the magnitudes such that $V_{L(p\text{-}p)} + V_{R(p\text{-}p)} = E_{p\text{-}p}$? If not, why not? How are they related?

TABLE 5.3

$V_{L(p\text{-}p)}$	$V_{R(p\text{-}p)}$	Sum

(j) Plot the curve of $I_{p\text{-}p}$ versus frequency on Graph 5.2. Label the curve and clearly indicate each plot point.

(k) How does the curve of $I_{p\text{-}p}$ versus frequency compare to the curve of $V_{R(p\text{-}p)}$ versus frequency? Explain why they compare as they do.

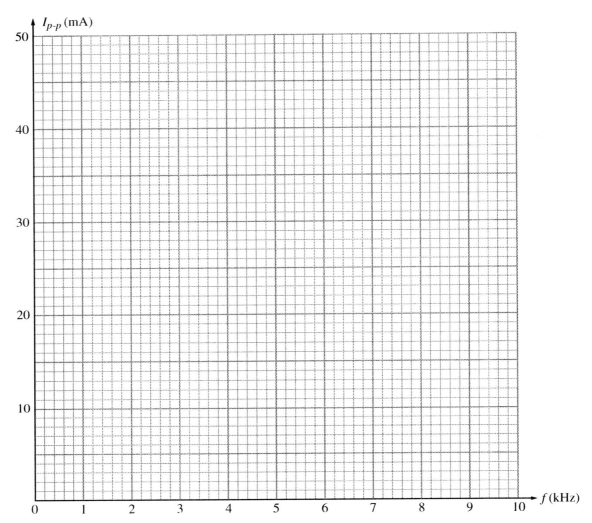

GRAPH. 5.2

(l) At a frequency of 8 kHz, calculate the reactance of the inductor using $X_L = 2\pi f L$ and the nameplate inductance level and record in Table 5.4. Record the value obtained from the data of Table 5.1 using

$$X_L = \frac{V_{L(p-p)}}{I_{p-p}}$$

Calculation:

TABLE 5.4

	Calculated	From Table 5.1 data
X_L		

Compare values of X_L.

(m) Use the Pythagorean theorem ($V_{L(p\text{-}p)} = \sqrt{E_{s(p\text{-}p)^2} - V_{R(p\text{-}p)^2}}$) to determine the voltage $V_{L(p\text{-}p)}$ at a frequency of 5 kHz and compare with the measured result of Table 5.1. Use the peak-to-peak value of V_R from Table 5.1 and $E_{p\text{-}p} = 4$ V. Insert the results in Table 5.5.

TABLE 5.5

	Calculated	Measured
$V_{L(p\text{-}p)}$		

(n) At low frequencies, the inductor approaches a low-impedance short-circuit equivalent and at high frequencies a high-impedance open-circuit equivalent. Do the data of Table 5.1 and Graphs 5.1 and 5.2 verify the above statement? Comment accordingly.

Part 2 Z_T versus Frequency

(a) For each frequency, transfer the results of $I_{p\text{-}p}$ from Table 5.1 to Table 5.6.

TABLE 5.6

Frequency	$E_{p\text{-}p}$	$I_{p\text{-}p}$	$Z_T = \dfrac{E_{p\text{-}p}}{I_{p\text{-}p}}$	$Z_T = \sqrt{R^2 + X_L^2}$
0.1 kHz	4 V			
1 kHz	4 V			
2 kHz	4 V			
3 kHz	4 V			
4 kHz	4 V			
5 kHz	4 V			

Continued

TABLE 5.6   Continued

Frequency	$E_{p\text{-}p}$	$I_{p\text{-}p}$	$Z_T = \dfrac{E_{p\text{-}p}}{I_{p\text{-}p}}$	$Z_T = \sqrt{R^2 + X_L^2}$
6 kHz	4 V			
7 kHz	4 V			
8 kHz	4 V			
9 kHz	4 V			
10 kHz	4 V			

(b) At each frequency, calculate the magnitude of the total impedance using the equation $Z_T = E_{p\text{-}p}/I_{p\text{-}p}$ in Table 5.6.

(c) Plot the curve of Z_T versus frequency on Graph 5.3. Calculate the total impedance at $f = 0$ Hz (with $R_l = 0 \ \Omega$) and include the result as a plot point for the curve. Label the curve and clearly indicate each plot point.

$$Z_T \, (f = 0 \text{ Hz}) = \underline{\hspace{3cm}}$$

(d) For each frequency, calculate the total impedance using the equation $Z_T = \sqrt{R^2 + X_L^2}$ and the measured value for R and insert in Table 5.6.

(e) How do the magnitudes of Z_T compare for the last two columns of Table 5.6?

(f) On the same graph (5.3), plot R versus frequency. Label the curve.

(g) On Graph 5.3, plot $X_L = 2\pi f L$ versus frequency. Use the space below for the necessary calculations. Label the curve and clearly indicate each plot point.

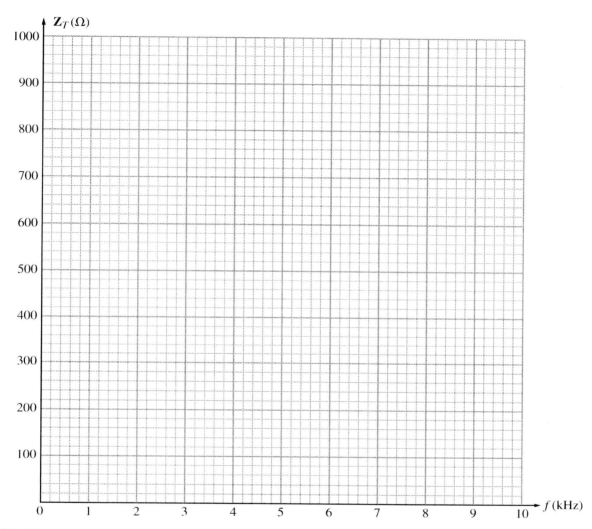

GRAPH. 5.3

(**h**) At which frequency does $X_L = R$? Use both the graph and a calculation ($f = R/2\pi L$). Record the results in Table 5.7.

Calculation:

How do the results compare?

TABLE 5.7

	Graph	Calculation
f		

(i) For frequencies less than the frequency calculated in part 2(h), is the network primarily resistive or inductive? How about for frequencies greater than the frequency calculated in part 2(h)?

(j) The phase angle by which the applied voltage leads the same current is determined by $\theta = \tan^{-1}(X_L/R)$ (as obtained from the impedance diagram). Calculate the phase angle for each of the frequencies in Table 5.8.

TABLE 5.8

Frequency	R (measured)	X_L	$\theta = \tan^{-1}(X_L/R)$
0.1 kHz			
1 kHz			
2 kHz			
3 kHz			
5 kHz			
10 kHz			
100 kHz			

(k) At a frequency of 0.1 kHz, does the phase angle suggest a primarily resistive or inductive network? Explain why.

(l) At frequencies greater than 5 kHz, does the phase angle suggest a primarily resistive or inductive network? Explain why.

(m) Plot θ versus frequency for the frequency range 0.1 kHz to 10 kHz on Graph 5.4. At what frequency is the phase angle equal to 45°? At what frequency is $X_L = R$? Using the relationship $X_L = R$, calculate the frequency at which $\theta = 45°$. Record the results in Table 5.9.

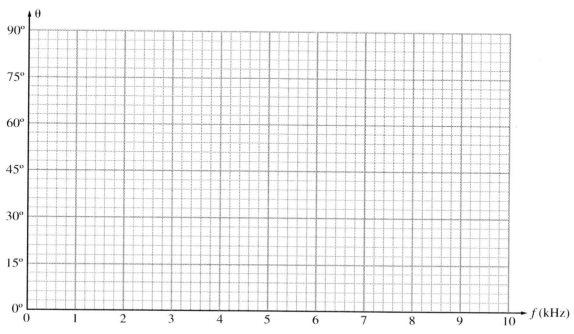

GRAPH. 5.4

TABLE 5.9

	$\theta = 45°$	$X_L = R$	Calculated
f			

How do the two levels of frequency compare?

EXERCISES

1. Given the network of Fig. 5.1 with $f = 1$ kHz, calculate the magnitude and phase angle of the input impedance and compare the results to those obtained experimentally in part 2(a) ($Z_T = E_{p\text{-}p}/I_{p\text{-}p}$) and calculated in Table 5.4.

Calculated:

$Z_T =$ _____

$\theta =$ _____

Experimental:

$Z_T =$ _____

$\theta =$ _____

2. Given the network of Fig. 5.1 with $f = 1$ kHz, calculate the levels of V_L, V_R, and I (all peak-to-peak values) and compare to the measured values of Table 5.2.

Calculated:

$V_L =$ _____

$V_R =$ _____

$I =$ _____

Measured:

$V_L =$ _____

$V_R =$ _____

$I =$ _____

3. Using PSpice or Multisim, obtain a plot of V_L, V_R and I versus frequency for the network of Fig. 5.1. Compare the results to that of Graph 5.1.

    Attach all appropriate printouts.

Frequency Response of the Series *R-C* Network

OBJECTIVES

1. Note the effect of frequency on the impedance of a series *R-C* network.
2. Plot the voltages and current of a series *R-C* network versus frequency.
3. Calculate and plot the phase angle of the input impedance versus frequency for a series *R-C* network.

EQUIPMENT REQUIRED

Resistors

1—1-kΩ (1/4-W)

Capacitors

1—0.1-μF

Instruments

1—DMM

1—Oscilloscope

1—Audio oscillator (or function generator)

1—Frequency counter (if available)

EQUIPMENT ISSUED

TABLE 6.0

Item	Manufacturer and Model No.	Laboratory Serial No.
DMM		
Oscilloscope		
Audio oscillator (or function generator)		
Frequency counter		

RÉSUMÉ OF THEORY

As noted in Experiment 5 for the series R-L network, the voltage across the coil increases with frequency since the inductive reactance increases directly with frequency and the impedance of the resistor is essentially independent of the applied frequency (in the audio range).

For the series R-C network, the voltage across the capacitor decreases with increasing frequency since the capacitive reactance is inversely proportional to the applied frequency.

Since the voltage and current of the resistor continue to be related by the fixed resistance value, the shapes of their curves versus frequency will have the same characteristics.

Again, keep in mind that the voltages across the elements in an ac circuit are vectorially related. Otherwise, the voltage readings may appear to be totally incorrect and not satisfy Kirchhoff's voltage law.

The phase angle associated with the input impedance is sensitive to the applied frequency. At very low frequencies, the capacitive reactance will be quite large compared to the series resistive element and the network will be primarily capacitive in nature. The result is a phase angle associated with the input impedance that approaches $-90°$ (v lags i by 90°). At increasing frequencies X_C will drop off in magnitude compared to the resistive element and the network will be primarily resistive, resulting in an input phase angle approaching 0° (v and i in phase).

Caution: **Be sure that the ground connections of the source and scope do not short out an element of the network, thereby changing its terminal characteristics.**

PROCEDURE

Part 1 V_C, V_R, and I versus Frequency

(a) Construct the network of Fig. 6.1. Insert the measured value of the resistor R on the diagram.

(b) Maintaining 4 V (p-p) at the input to the circuit, record the voltage $V_{C(p\text{-}p)}$ for the frequencies appearing in Table 6.1. Make sure to check continually that $E_s = 4$ V (p-p) with each frequency change. Do not measure the voltage V_R at this point in the experiment. The common

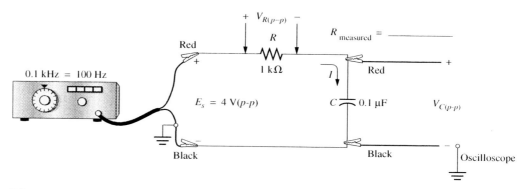

FIG. 6.1

ground of the supply and scope will short out the effect of the capacitive element, which may re-sult in damage to the equipment.

For each frequency, try to read V_C to the highest degree of accuracy possible. The higher the degree of accuracy, the better the data will verify the theory to be substantiated.

TABLE 6.1

Frequency	$V_{C(p\text{-}p)}$	$V_{R(p\text{-}p)}$	$I_{p\text{-}p}$
0.1 kHz			
0.2 kHz			
0.5 kHz			
1 kHz			
2 kHz			
4 kHz			
6 kHz			
8 kHz			
10 kHz			

(c) Turn off the supply and interchange the positions of R and C in Fig. 6.1 and mea-sure $V_{R(p\text{-}p)}$ for the same range of frequencies with E maintained at 4 V (p-p). Insert the measure-ments in Table 6.1. As in Experiment 5, it is vitally important that this step be performed as specified, or a grounding problem can result.

(d) Calculate $I_{p\text{-}p}$ from $I_{p\text{-}p} = V_{R(p\text{-}p)}/R_{\text{measured}}$ and complete Table 6.1.

(e) Plot the curve of $V_{C(p\text{-}p)}$ versus frequency on Graph 6.1. Label the curve and clearly indicate each plot point.

(f) Plot the curve of $V_{R(p\text{-}p)}$ versus frequency on Graph 6.1. Again, label the curve and clearly indicate each plot point.

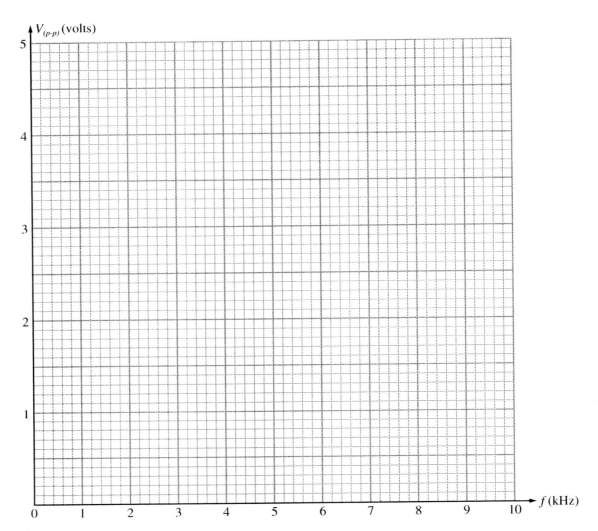

GRAPH 6.1

(g) As the frequency increases, describe in a few sentences what happens to the voltage across the capacitor and resistor. Explain why.

(h) At the point where $V_C = V_R$, does $X_C = R$? Should they be equal? Why? Record the level of voltage and the impedance of each element in Table 6.2.

TABLE 6.2

$V(V_C = V_R)$	X_C	R

(i) Determine $V_{C(p\text{-}p)}$ and $V_{R(p\text{-}p)}$ at some random frequency such as 3.6 kHz from the curves and record in Table 6.3. Determine the sum of their magnitudes and record in Table 6.3.

TABLE 6.3

$V_{C(p\text{-}p)}$	$V_{R(p\text{-}p)}$	Sum

Are the magnitudes such that $V_{C(p\text{-}p)} + V_{R(p\text{-}p)} = E_{p\text{-}p}$?

If not, why not? How are they related?

(j) Plot the curve of $I_{p\text{-}p}$ versus frequency on Graph 6.2. Label the curve and clearly indicate each plot point.

(k) How does the curve of $I_{p\text{-}p}$ versus frequency compare to the curve of $V_{R(p\text{-}p)}$ versus frequency? Explain why they compare as they do.

(l) At a frequency of 6 kHz, calculate the reactance of the capacitor using $X_C = 1/(2\pi fC)$ and the nameplate capacitance level and record in Table 6.4. Then determine the value obtained from the data of Table 6.1 using

$$X_C = \frac{V_{C(p\text{-}p)}}{I_{p\text{-}p}}$$

and record in Table 6.4.

Calculation:

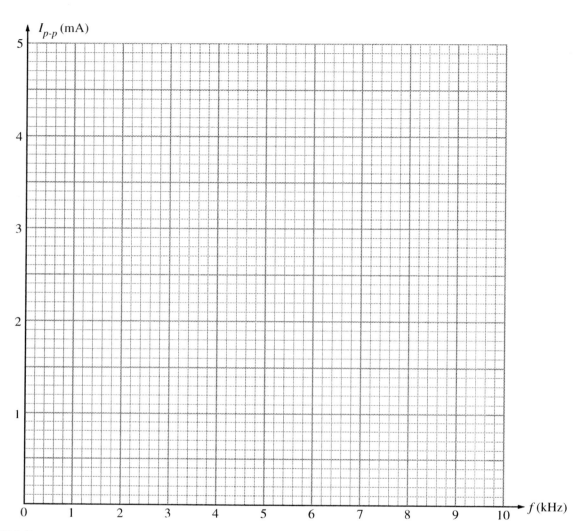

GRAPH 6.2

TABLE 6.4

	Calculated	Table 6.2
X_C		

Compare the values determined in Table 6.4.

(m) Use the Pythagorean theorem to determine the voltage $V_{C(p-p)}$ at a frequency of 6 kHz ($V_{C(p-p)} = \sqrt{E_{s(p-p)}^2 - V_{R(p-p)}^2}$) and compare with the measured result of Table 6.1. Use the peak-to-peak value of V_R from Table 6.1 and $E_{s(p-p)} = 4$ V. Record both results in Table 6.5.

Calculation:

(n) At low frequencies, the capacitor approaches a high-impedance open-circuit equivalent, and at high frequencies, a low-impedance short-circuit equivalent. Do the data of Table 6.1 and Graphs 6.1 and 6.2 verify the above statement? Comment accordingly.

TABLE 6.5

	Calculated	Table 6.1
$V_{C(p-p)}$		

Part 2 Z_T versus Frequency

(a) For each frequency, transfer the results of I_{p-p} from Table 6.1 to Table 6.6.

TABLE 6.6

Frequency	E_{p-p}	I_{p-p}	$Z_T = \dfrac{E_{p-p}}{I_{p-p}}$	$Z_T = \sqrt{R^2 + X_C^2}$
0.1 kHz	4 V			
0.2 kHz	4 V			
0.5 kHz	4 V			
1 kHz	4 V			
2 kHz	4 V			
4 kHz	4 V			
6 kHz	4 V			
8 kHz	4 V			
10 kHz	4 V			

(b) At each frequency, calculate the magnitude of the total impedance using the equation $Z_T = E_{p-p}/I_{p-p}$ in Table 6.6.

(c) Plot the curve of Z_T versus frequency on Graph 6.3 except for $f = 0.1$ kHz, which is off the graph. Label the curve and clearly indicate each plot point.

(d) For each frequency, calculate the total impedance using the equation $Z_T = \sqrt{R^2 + X_C^2}$ and insert in Table 6.6.

(e) How do the magnitudes of Z_T compare for the last two columns of Table 6.6?

(f) On Graph 6.3, plot R versus frequency. Label the curve.

(g) On Graph 6.3, plot $X_C = 1/2\pi fC$ versus frequency. Label the curve and clearly indicate each plot point.

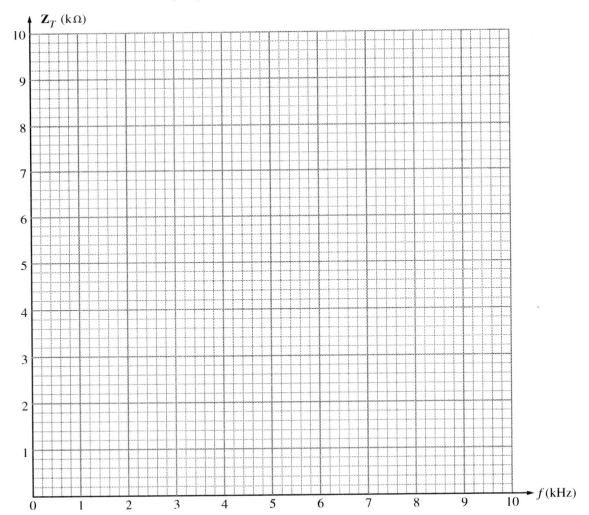

GRAPH 6.3

(h) At which frequency does $X_C = R$? Use both the graph and a calculation ($f = 1/2\pi RC$). How do they compare? Record both results in Table 6.7.

Calculation:

TABLE 6.7

	Graph	Calculation
f		

(i) For frequencies less than the frequency calculated in part 2(h), is the network primarily resistive or capacitive? How about for frequencies greater than the frequency calculated in part 2(h)?

(j) The phase angle by which the applied voltage leads the current is determined by $\theta = -\tan^{-1}(X_C/R)$ (as obtained from the impedance diagram). The negative sign is a clear indication that for capacitive networks, i leads v. Determine the phase angle for each of the frequencies in Table 6.8.

TABLE 6.8

Frequency	R (measured)	X_C	$\theta = -\tan^{-1}(X_C/R)$
0.1 kHz			
0.2 kHz			
0.5 kHz			
1 kHz			
2 kHz			
6 kHz			
10 kHz			
100 kHz			

(k) At a frequency of 0.1 kHz, does the phase angle suggest a primarily resistive or capacitive network? Explain why.

(l) At frequencies greater than 2 kHz, does the phase angle suggest a primarily resistive or capacitive network? Explain why.

(m) Plot θ versus frequency for the frequency range 0.2 kHz to 10 kHz on Graph 6.4. At what frequency is the phase angle equal to $-45°$? At what frequency does $X_C = R$? Using the relationship $X_C = R$ calculate the frequency at which $\theta = -45°$. Record the results in Table 6.9.

TABLE 6.9

	$\theta = -45°$	$X_C = R$	**Calculated**
f			

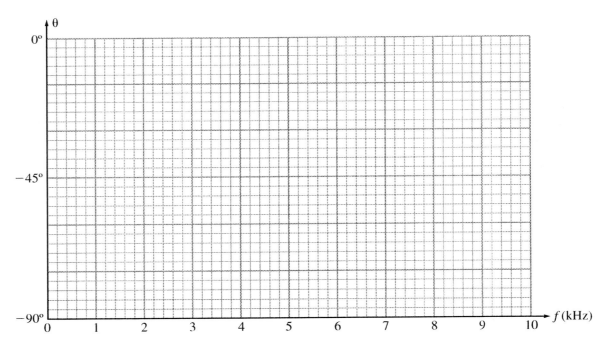

GRAPH 6.4

How do the levels of frequency compare?

EXERCISES

1. Given the network of Fig. 6.1, with $f = 1$ kHz, calculate the magnitude and phase angle of the input impedance and compare the results to those obtained experimentally in part 2(a) ($Z_T = E_{p-p}/I_{p-p}$) and calculated in Table 6.8.

Calculated: Experimental:

$Z_T =$ _____ $Z_T =$ _____

$\theta =$ _____ $\theta =$ _____

2. Given the network of Fig. 6.1 with $f = 1$ kHz, calculate the levels of V_C, V_R, and I (all peak-to-peak values) and compare to the measured values of Table 6.1.

Calculated: Measured:

$V_C =$ _____ $V_C =$ _____

$V_R =$ _____ $V_R =$ _____

$I =$ _____ $I =$ _____

3. Using PSpice or Multisim, obtain a plot of V_C, V_R, and I versus frequency for the network of Fig. 6.1. Compare the results to that of Graph 6.1.

      Attach all appropriate printouts.

The Oscilloscope and Phase Measurements

OBJECTIVES

1. Use the dual trace (two vertical channels) to determine the phase angle between two sinusoidal waveforms.
2. Become aware of the use of Lissajous patterns to determine the phase angle between two sinusoidal waveforms.
3. Become aware of the effect of increasing levels of resistance on the phase angles of a series *R-C* circuit.

EQUIPMENT REQUIRED

Resistors

1—1-kΩ, 3.3-kΩ, 6.8-kΩ (1/4-W)

Capacitors

1—0.47-μF

Instruments

1—DMM
1—Oscilloscope
1—Audio oscillator or function generator
1—Frequency counter (if available)

EQUIPMENT ISSUED

TABLE 7.0

Item	Manufacturer and Model No.	Laboratory Serial No.
DMM		
Oscilloscope		
Audio oscillator or function generator		
Frequency counter		

RÉSUMÉ OF THEORY

The phase angle between two signals of the same frequency can be determined using the oscilloscope. There are two methods available.

1. Dual-trace comparison with the calibrated time base
2. Lissajous pattern

Dual-Trace Method of Phase Measurement

The dual-trace method of phase measurement, aside from providing a high degree of accuracy, can compare two signals of different amplitudes and, in fact, different wave shapes. The method can be applied directly to oscilloscopes equipped with two vertical channels or to a conventional single-trace oscilloscope with an external electronic switch, as shown in Fig. 7.1. The electronic switch will switch between inputs at a very high speed, so both patterns will appear on the screen.

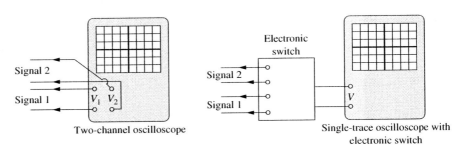

FIG. 7.1

Regardless of which oscilloscope is available, the procedure essentially consists of displaying both traces on the screen simultaneously and measuring the distance (in scale divisions) between two identical points on the two traces (Fig. 7.2).

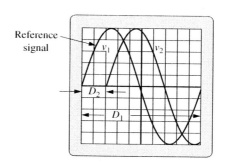

Reference signal

FIG. 7.2

One signal will be chosen as a reference, that is, zero-phase angle. In the comparison, therefore, we can assume that the signal being compared is leading $(+\theta)$ if it is to the left of the reference and lagging $(-\theta)$ if it is to the right of the reference. To use the dual-trace phase measurement method, therefore, proceed as follows:

1. Connect the two signals to the two vertical channels, making sure to observe proper grounding. For clarity, adjust the vertical sensitivity of each waveform until both signals have the same relative size.
2. Select the mode of operation—"Alternate" or "Chop." For frequencies less than 50 kHz, use Chop. For frequencies greater than 50 kHz, use Alternate.
3. Once the traces are on the screen, use the GND switch to set both patterns in the vertical center of the screen.
4. Measure the number of horizontal divisions (D_1 in Fig. 7.2) required for one full cycle of either waveform (they both have the same frequency).
5. Measure the number of horizontal divisions in the phase shift (D_2), as shown in Fig. 7.2.
6. Because D_1 is associated with a full cycle of 360° and D_2 is associated with the phase angle θ, we can set up the following ratio and solve for θ:

$$\frac{D_1}{360°} = \frac{D_2}{\theta} \qquad \textbf{(7.1)}$$

or

$$\theta = \frac{D_2}{D_1} \times 360° \qquad \textbf{(7.2)}$$

For the case of Fig. 7.2,

$$\theta = \frac{2 \text{ div.}}{10 \text{ div.}} \times 360° = \textbf{72°}$$

Lissajous-Pattern Phase Measurement

The Lissajous-pattern method is also called the *X-Y* phase measurement. To use this method, proceed as follows:

1. Connect one signal to a vertical channel and the other to the horizontal input (often denoted *X-Y*).

2. A display known as a *Lissajous* pattern will appear on the screen. The type of pattern will reveal the phase relationship, and, in fact, the pattern can be used to calculate the phase angle. It will define the angle by which the horizontal input leads the vertical input.

The patterns shown in Fig. 7.3 indicate the phase relationship appearing with each figure.

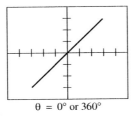 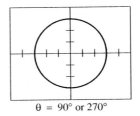 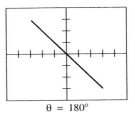

$\theta = 0°$ or $360°$ $\theta = 90°$ or $270°$ $\theta = 180°$

FIG. 7.3

The patterns shown in Fig. 7.4 can be used to calculate the phase angle (θ), as indicated below the figures.

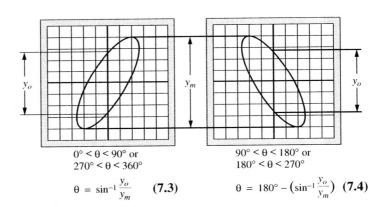

$0° < \theta < 90°$ or $90° < \theta < 180°$ or
$270° < \theta < 360°$ $180° < \theta < 270°$

$$\theta = \sin^{-1}\frac{y_o}{y_m} \quad (7.3)$$ $$\theta = 180° - \left(\sin^{-1}\frac{y_o}{y_m}\right) \quad (7.4)$$

FIG. 7.4

EXAMPLE   Assume that the patterns in Figs. 7.5 and 7.6 appear on an oscilloscope screen. Calculate the phase angle θ in each case.

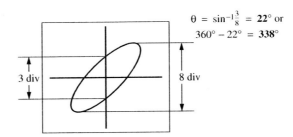

$$\theta = \sin^{-1}\tfrac{3}{8} = \mathbf{22°} \text{ or}$$
$$360° - 22° = \mathbf{338°}$$

3 div   8 div

FIG. 7.5

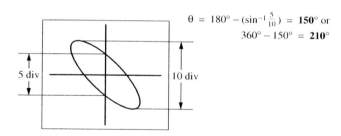

$$\theta = 180° - (\sin^{-1}\tfrac{5}{10}) = \mathbf{150°} \text{ or}$$
$$360° - 150° = \mathbf{210°}$$

FIG. 7.6

PROCEDURE

Part 1

The phase relationship between **E** and $\mathbf{V}_R$ of Fig. 7.7 will be determined in this part using the dual-trace capability of the oscilloscope.

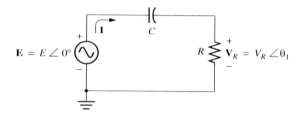

FIG. 7.7

Since **E** is defined as having an angle of 0°, it will appear as shown in Fig. 7.8. In an *R-C* circuit, the current **I** will lead the applied voltage as shown in the phasor diagram. The voltage $\mathbf{V}_R$ is in phase with **I**, and the voltage $\mathbf{V}_C$ will lag the voltage **E**.

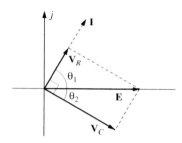

FIG. 7.8

Note that $|\theta_1| + |\theta_2| = 90°$ and that the vector sum of $\mathbf{V}_R$ and $\mathbf{V}_C$ equals the applied voltage **E**.

(a) Construct the network of Fig. 7.9, designed to place the source voltage **E** on channel 1 of the scope and $\mathbf{V}_R$ on channel 2.

The oscillator or function generator is set to 200 Hz with an amplitude of 8 V (p-p). Set the vertical and horizontal sensitivities as indicated on Fig. 7.9.

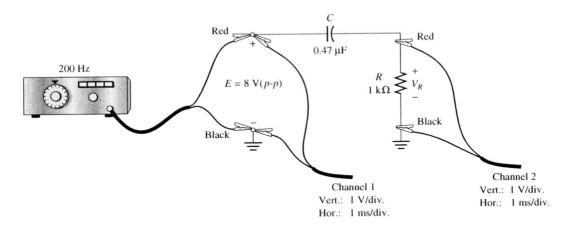

FIG. 7.9

(b) Enter the measured value of R in Table 7.1. Then calculate the reactance of the capacitor or a frequency of 200 Hz and enter in the second column of Table 7.1.

TABLE 7.1

R	X_C	$E_{(p-p)}$	$V_{R_{(p-p)}}$		θ_1		D_1	D_2
			Calculated	**Measured**	**Calculated**	**Measured**		
		8 V						

(c) Assuming $\mathbf{E} = 8$ V (p-p) $\angle 0°$, calculate the peak-to-peak value of $\mathbf{V}_R$ and the angle (θ_1) associated with $\mathbf{V}_R$ and insert in Table 7.1. Show all work! Be precise.

Calculation:

(d) Energize the network of Fig. 7.9 and determine the peak-to-peak value of V_R. Insert in the "Measured" column of Table 7.1.

(e) Determine the number of horizontal divisions for one full cycle of **E** or V_R. Enter as D_1 in Table 7.1.

Determine the number of horizontal divisions representing the phase shift between waveforms. Enter as D_2 in Table 7.1.

Determine the phase shift in degrees using Eq. (7.2). Enter as the "Measured" value in Table 7.1.

Calculation:

(f) How do the calculated and measured values for V_R and θ_1 compare?

(g) Replace the 1-kΩ resistor with a 3.3-kΩ resistor and repeat the above calculations and measurements. Show all calculations below and insert the results in Table 7.2. Be neat.

Calculation:

TABLE 7.2

R	X_C	$E_{(p-p)}$	$V_{R_{(p-p)}}$		θ_1	
			Calculated	Measured	Calculated	Measured
		8 V				

(h) Replace the 3.3-kΩ resistor with a 6.8-kΩ resistor and repeat the above calculations and measurements. Show all calculations below and insert the results in Table 7.3. Be neat.

Calculation:

TABLE 7.3

R	X_C	$E_{(p-p)}$	$V_{R_{(p-p)}}$		θ_1	
			Calculated	Measured	Calculated	Measured
		8 V				

(i) Are you satisfied with the results of parts (g) and (h)? Comment accordingly.

(j) The vector $\mathbf{E} = E \angle\theta = 8$ V $\angle 0°$ has been placed on each phasor diagram of Graph 7.1. Note that the voltage has been scaled to match the 2-V/div. scale of the horizontal and vertical axes. Using the measured values of $V_{R_{(p-p)}}$ and θ_1, insert the phasor $\mathbf{V}_R$ for each value of R. Clearly indicate the angle θ_1 and the magnitude of $V_{R_{(p-p)}}$.

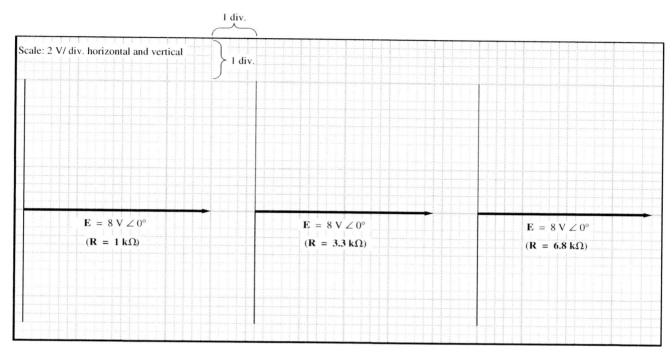

GRAPH 7.1

Part 2

The phase relationship between **E** and V_C of the same network as Fig. 7.7 will now be determined by interchanging the resistor R and capacitor C, as shown in Fig. 7.10.

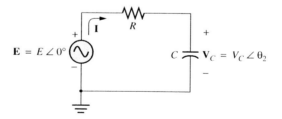

FIG. 7.10

If the oscilloscope were simply placed across the capacitor of Fig. 7.7, the elements must exchange positions to avoid a "shorting out" of the resistor R of Fig. 7.7. The grounds of the scope and supply would establish a 0-V drop across R and possibly high currents in the remaining network, since X_C is the only impedance to limit the current level.

(a) The connections are now made as shown in Fig. 7.11. Note that **E** remains on channel 1 and V_C is placed on channel 2. The phase angle θ_2 between **E** and V_C can therefore be determined from the display.

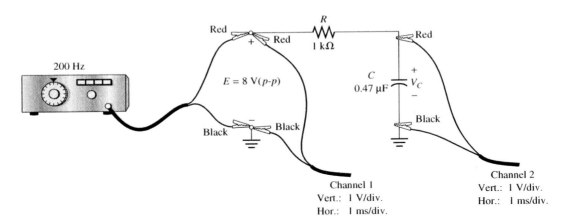

FIG. 7.11

(b) Enter the measured value of R in Table 7.4. Then enter the reactance at 200 Hz from Table 7.1.

TABLE 7.4

R	X_C	$E_{(p\text{-}p)}$	$V_{C_{(p-p)}}$		θ_2		D_1	D_2
			Calculated	**Measured**	**Calculated**	**Measured**		
		8 V						

(c) Assuming $\mathbf{E} = 8$ V (p-p) $\angle 0°$, calculate the peak-to-peak value of $\mathbf{V}_C$ and the angle (θ_2) associated with $\mathbf{V}_C$ and insert in Table 7.4. Show all work! Be neat.

Calculation:

(d) Energize the network of Fig. 7.11 and determine the peak-to-peak value of $\mathbf{V}_C$. Insert in the measured column of Table 7.4.

(e) Determine the phase shift in the same manner as described for part 1. Record D_1 and D_2 in Table 7.4. Record the value for θ_2 in the measured column of Table 7.4.

Calculation:

(f) How do the calculated and measured values for V_C and θ_2 compare?

(g) Replace the 1-kΩ resistor by a 3.3-kΩ resistor and repeat the above calculations and measurements. Show all calculations below and insert the results in Table 7.5. Be neat.

Calculation:

TABLE 7.5

R	X_C	$E_{(p\text{-}p)}$	$V_{C_{(p\text{-}p)}}$		θ_2	
			Calculated	Measured	Calculated	Measured
		8 V				

(h) Replace the 3.3-kΩ resistor with a 6.8-kΩ resistor and repeat the above calculations and measurements. Show all calculations below and insert the results in Table 7.6. Be neat!

Calculation:

TABLE 7.6

R	X_C	$E_{(p\text{-}p)}$	$V_{C_{(p\text{-}p)}}$		θ_2	
			Calculated	Measured	Calculated	Measured
		8 V				

(i) Are you satisfied with the results of parts (g) and (h)? Comment accordingly.

(j) Complete the phasor diagrams of Graph 7.1 with the insertion of the phasor V_C for each resistance level using peak-to peak values for the voltages. Label each vector.

(k) It was noted in Fig. 7.8 that $|\theta_1| + |\theta_2| = 90°$. For each resistance level, add the two measured values and determine the magnitude of the percent difference using the equation

$$\% \text{ Difference} = \left| \frac{90° - \theta_T}{90°} \right| \times 100\% \qquad (7.5)$$

Record the values in Table 7.7.

TABLE 7.7

R	θ_1 (meas., part 1)	θ_2 (meas., part 2)	$|\theta_T| = |\theta_1| + |\theta_2|$	% Difference 90° vs. θ_T
1 kΩ				
3.3 kΩ				
6.8 kΩ				

Part 3

In this part, the phase angle will be determined from the Lissajous pattern.

(a) Construct the network of Fig. 7.12, which is exactly the same as Fig. 7.9 except now V_R is connected to the horizontal input.

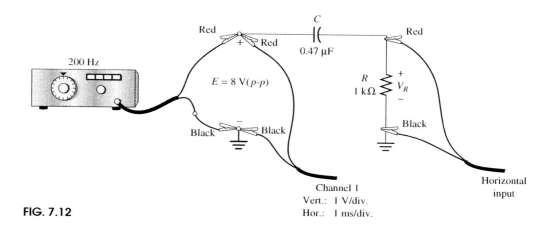

FIG. 7.12

The resulting Lissajous patterns will determine the phase angle between **E** and $\mathbf{V}_R$ for various values of R as described in the Résumé of Theory.

(b) Change the value of R as indicated in Table 7.8 and measure the y-intercept (y_o) and the y-maximum (y_m) from the Lissajous pattern. Record in Table 7.8.

TABLE 7.8

R	y_o	y_m	θ_1
1 kΩ			
3.3 kΩ			
6.8 kΩ			

(c) Then compare the measured values of θ_1 obtained in Tables 7.1, 7.2, and 7.3 with those in Table 7.8 by completing Table 7.9, where

$$\% \text{ Difference} = \left| \frac{\theta_1(\text{Tables 7.1, 7.2, 7.3}) - \theta_1(\text{Table 7.8})}{\theta_1(\text{Tables 7.1, 7.2, 7.3})} \right| \times 100\% \qquad \textbf{(7.6)}$$

TABLE 7.9

R	θ_1 (Tables 7.1, 7.2, 7.3)	θ_1 (Table 7.8)	% Difference
1 kΩ			
3.3 kΩ			
6.8 kΩ			

(d) Which method do you believe provided the higher degree of accuracy for determining θ_1? With which method are you more comfortable? Why?

PROBLEMS

1. Determine the phase shift between the two sinusoidal voltages of Fig. 7.13.

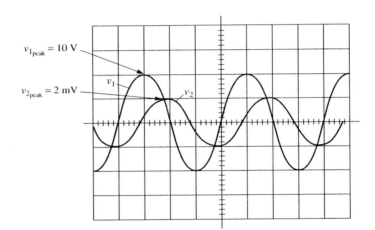

FIG. 7.13

$\theta = $ _____

2. Determine the phase shift between two sinusoidal waveforms that established the Lissajous pattern of Fig. 7.14.

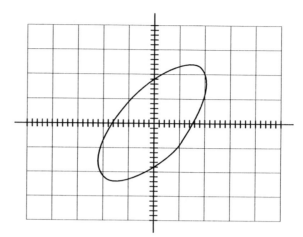

FIG. 7.14

$\theta = $ _____

Name:_____

Date:_____

Course and Section:_____

Instructor:_____

Series Sinusoidal Circuits

OBJECTIVES

1. Measure the voltages of a series ac circuit and verify Kirchhoff's voltage law.
2. Determine the input impedance using experimental methods.
3. Apply the dual-trace method to determine the phase angle associated with each voltage of the circuit.
4. Verify the voltage divider rule through actual measurements.
5. Analyze a network with all three elements, R, L, and C.

EQUIPMENT REQUIRED

Resistors

1—1-kΩ (1/4-W)

Inductors

1—10-mH

Capacitors

1—0.01-μF

Instruments

1—DMM

1—Oscilloscope

1—Audio oscillator or function generator

1—Frequency counter (if available)

EQUIPMENT ISSUED

TABLE 8.0

Item	Manufacturer and Model No.	Laboratory Serial No.
DMM		
Oscilloscope		
Audio oscillator or function generator		
Frequency counter		

RÉSUMÉ OF THEORY

Kirchhoff's voltage law is applicable to ac circuits, but it is now stated as follows: *The phasor sum of the voltages around a closed loop is equal to zero.*

For example, in a series circuit, the source voltage is the phasor sum of the component (load) voltages. For a series *R-L-C* circuit,

$$E^2 = V_R^2 + (V_L - V_C)^2$$

so that

$$E = \sqrt{V_R^2 + (V_L - V_C)^2} \qquad (8.1)$$

where E is the source voltage, V_R is the voltage across the total resistance of the circuit, V_L is the voltage across the total inductance, and V_C is the voltage across the total capacitance.

Since the resistance and the reactance of a series *R-L-C* circuit are in quadrature,

$$Z^2 = R^2 + X_T^2 \qquad (8.2)$$

where X_T (total reactance) $= X_L - X_C$.

In a series *R-L-C* sinusoidal circuit, the voltage across a reactive component may be greater than the input voltage.

For an ideal inductor, the current lags the voltage across it by 90°. For a capacitor, the current leads the voltage across it by 90°. Inductive circuits are therefore called *lagging power-factor circuits,* and capacitive circuits are called *leading power-factor circuits.*

In a purely resistive circuit, the voltage and the current are in phase and the power factor is unity.

PROCEDURE

Part 1   Series *R-L* Circuit

(a) Construct the network of Figure 8.1. Insert the measured values of *R*. The internal dc resistance of the coil will be ignored because it is so small compared to the other series elements of the circuit.

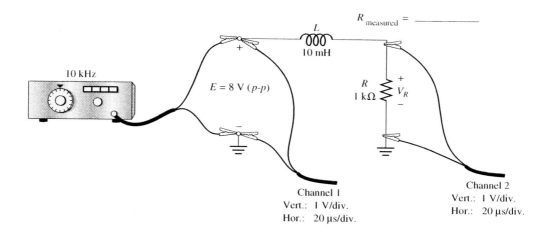

FIG. 8.1

(b) After setting E to 8 V $(p\text{-}p)$, determine the peak-to-peak voltage for V_R from channel 2 and record in the top row of Table 8.1.

(c) Determine the phase angle θ_1 between $\mathbf{E}$ and $\mathbf{V}_R$ using the connections shown in Fig. 8.1 and the dual-trace method introduced in Experiment 7. Record the results in the top row of Table 8.1.

Calculation:

TABLE 8.1

	$V_{R_{(p\text{-}p)}}$	D_1	D_2	θ (measured)	θ (calculated)
R and θ_1					
R and θ_2					

(d) Determine $I_{p\text{-}p}$ from $I_{p\text{-}p} = V_{R(p\text{-}p)}/R_{\text{measured}}$. Enter the result in the first column of Table 8.2.

Calculation:

(e) Determine the input impedance from $Z_T = E_{p\text{-}p}/I_{p\text{-}p}$. Enter the results in the first column of Table 8.2.

Calculation:

(f) Calculate the total impedance (magnitude and angle) and current $I_{p\text{-}p}$ from the nameplate value of the inductance and the measured resistor value. Ignore the effect of R_l. Enter the results in the second column of Table 8.2.

Calculation:

(g) Compare the results of (e) and (f) and explain the source of any differences.

The next step is important for grounding; reasons will be explained by your instructor!

(h) Reverse the positions of R and L and measure $V_{L_{(p\text{-}p)}}$ (to ensure that the oscillator (or generator) and the oscilloscope have a common ground). Record this value in the second row of Table 8.1.

(i) Determine the phase angle θ_2 between $\mathbf{E}$ and $\mathbf{V}_L$ using the dual-trace method introduced in Experiment 7. Record the results in the second row of Table 8.1.

Calculation:

(**j**) Show that your measurements from parts (b) and (h) satisfy Kirchhoff's voltage law. That is, show that $E = \sqrt{V_R^2 + V_L^2}$ using peak-to-peak values.

Calculation:

(**k**) The phase angle between $\mathbf{V}_R$ (in phase with $\mathbf{I}$) and $\mathbf{V}_L$ should be 90°. Does $|\theta_1| + |\theta_2| = 90°$? What is the magnitude of the percent difference $(|(90° - \text{sum})|/90°) \times 100\%$?

% Difference = _____

(**l**) Using $\mathbf{E} = 8\ \text{V} \angle 0°$, $R = 1\ \text{k}\Omega$, and $L = 10\ \text{mH}$, *calculate* $\mathbf{V}_R$, $\mathbf{V}_L$, and $\mathbf{I}$ at $f = 10\ \text{kHz}$ (peak-to-peak values) and draw the phasor diagram using peak-to-peak values. Determine the phase angle θ_1 between $\mathbf{E}$ and $\mathbf{V}_R$ and compare to the measured value of (c). Determine θ_2 from the phasor diagram and compare to θ_2 of (i). Record the values of θ_1 and θ_2 in Table 8.1. Show all work. Organize the presentation.

Calculation:

(**m**) Complete Table 8.2.

TABLE 8.2

Quantity	Measured (or Calculated from Measured Values)	Theoretical (Calculated)
$E_{p\text{-}p}$		
$V_{R_{(p\text{-}p)}}$		
$V_{L_{(p\text{-}p)}}$		
$I_{p\text{-}p}$		
Z_T		
θ_T		

(**n**) Compare the measured and theoretical results and try to explain any major differences.

Part 2   Series *R-C* Circuit

(**a**) Construct the network of Fig. 8.2. Insert the measured resistor value.

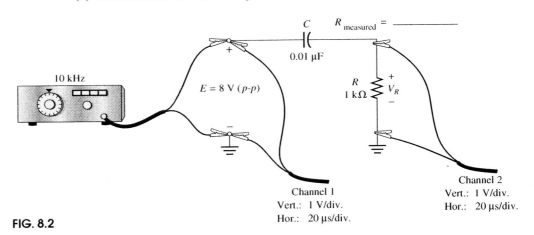

FIG. 8.2

(**b**) After setting *E* to 8 V (*p-p*), measure the voltage $V_{R_{(p\text{-}p)}}$ with the oscilloscope and record in the top row of Table 8.3.

TABLE 8.3

	$V_{R_{(p\ p)}}$	D_1	D_2	θ (measured)	θ (calculated)
R and θ_1					
C and θ_2					

(c) Determine the phase angle θ_1 between **E** and $\mathbf{V}_R$ using the connections shown in Fig. 8.2 and the dual-trace method introduced in Experiment ac 7. Record the results in the top row of Table 8.3.

Calculation:

(d) Determine $I_{p\text{-}p}$ from $I_{p\text{-}p} = V_{R(p\text{-}p)}/R_{\text{measured}}$. Enter the results in the first column of Table 8.4.

Calculation:

(e) Determine the input impedance from $Z_T = E_{p\text{-}p}/I_{p\text{-}p}$ and record in the first column of Table 8.4.

Calculation:

(f) Calculate the total impedance (magnitude and angle) and current $I_{p\text{-}p}$ from the nameplate value of the capacitance and the measured resistor value. Record the results in the second column of Table 8.4.

Calculation:

(g) Compare the results of (e) and (f) and explain the source of any differences.

(h) Reverse the positions of R and C and measure $V_{C(p \cdot p)}$ (to ensure that the oscilla-tor (or generator) and the oscilloscope have a common ground). Record the results in the bottom row of Table 8.3.

(i) Determine the phase angle θ_2 between $\mathbf{E}$ and $\mathbf{V}_C$ using the dual-trace method in-troduced in Experiment 7 and record the results in the bottom row of Table 8.3.

Calculation:

(j) The phase angle between $\mathbf{V}_R$ (in phase with $\mathbf{I}$) and $\mathbf{V}_C$ should be 90°. Does $|\theta_1| + |\theta_2| = 90°$? What is the magnitude of the percent difference $(|90° - \text{sum}|/90°) \times 100\%$?

% Difference = _____

(k) Using $\mathbf{E} = 8$ V $\angle 0°$, $R = 1$ kΩ, and $C = 0.01$ μF, *calculate* $\mathbf{V}_R$, $\mathbf{V}_C$, and $\mathbf{I}$ at $f = 10$ kHz (peak-to-peak values) and draw the phasor diagram using peak-to-peak values. Determine the phase angle θ_1 between $\mathbf{E}$ and $\mathbf{V}_R$ and compare to the measured value of 2(c). Determine θ_2 from the phasor diagram and compare to θ_2 of (i). Record the results for θ_1 and θ_2 in last column of Table 8.3.

Calculation:

(**l**) Complete Table 8.4.

TABLE 8.4

Quantity	Measured (or Calculated from Measured Values)	Theoretical (Calculated)
$E_{p\text{-}p}$		
$V_{R(p\text{-}p)}$		
$V_{C(p\text{-}p)}$		
$I_{p\text{-}p}$		
Z_T		
θ_T		

(**m**) Compare the measured and theoretical results and try to explain any major differences.

Part 3   Series *R-L-C* Circuit

(**a**) Construct the network of Figure 8.3. Insert the measured resistance values. Ignore the effects of R_l in the following analysis.

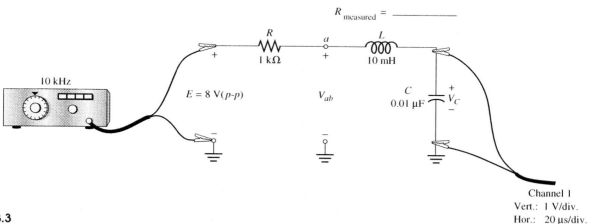

FIG. 8.3

(**b**) Measure all the component voltages with $E = 8$ V (*p-p*). To ensure a common ground between the oscilloscope and oscillator for each measurement, make sure the element is placed in the position of the capacitor *C*. In other words, reverse the order of the elements so that the element across which the voltage is to be measured has the position of *C* indicated in Fig. 8.3. Record the results in Table 8.5.

TABLE 8.5

	Measured	Calculated
$V_{R(p\text{-}p)}$		
$V_{L(p\text{-}p)}$		
$V_{C(p\text{-}p)}$		
$I_{(p\text{-}p)}$		
Z_T		
$V_{ab(p\text{-}p)}$		

(c) Determine $I_{p\text{-}p}$ from $I_{p\text{-}p} = V_{R(p\text{-}p)}/R_{\text{measured}}$ and record in the first column of Table 8.5.

Calculation:

(d) Calculate Z_T from $Z_T = E_{p\text{-}p}/I_{p\text{-}p}$ and place the result in the first column of Table 8.5.

Calculation:

(e) Using the nameplate values for L and C and the measured value for R, calculate Z_T and record in the second column of Table 8.5.

Calculation:

Compare with the result from part 3(d).

(f) Using $\mathbf{E} = 8\,\text{V}\angle 0°$, find $\mathbf{I}$, $\mathbf{V}_R$, $\mathbf{V}_L$, and $\mathbf{V}_C$ using peak-to-peak values. Record the results in the second column of Table 8.5.

Calculation:

(g) Draw the phasor diagram, including $\mathbf{I}$ and all the voltages.

(h) Verify Kirchhoff's law by showing that $E = \sqrt{V_R^2 + (V_L - V_C)^2}$ using the measured peak-to-peak values.

(i) Use the voltage divider rule to calculate the voltage $V_{ab(p\text{-}p)}$ and record in the second column of Table 8.5.

Calculation:

(j) Measure the voltage V_{ab} and record in Table 8.5. Compare to the result of 3(h).

EXERCISES

1. Using measured voltage levels and $R_{measured}$, determine the actual inductance L of Fig. 8.1 at a frequency of 10 kHz.

   Calculation:

   $L = $ _____

2. Using measured voltage levels and $R_{measured}$, determine the actual capacitance C of Fig. 8.2 at a frequency of 10 kHz.

Calculation:

$C =$ _____

3. Using the inductance L from problem 1 and the capacitance C from problem 2, calculate the total impedance Z_T for the network of Fig. 8.3 and compare to the measured value.

Calculation:

$Z_T =$ _____

4. Using PSpice or Multisim, determine the voltages V_R, V_L, and V_C for the network of Fig. 8.3 and compare with the measured values of Table 8.5. Attach all appropriate printouts.

EXPERIMENT ac **9**

Parallel Sinusoidal Circuits

OBJECTIVES

1. Become aware of the usefulness of sensing resistors to determine current levels and associated phase angles.
2. Determine the input impedance using experimental methods.
3. Apply the dual-trace method to determine the phase angle associated with each current of the network.
4. Verify Kirchhoff's current law through actual measurements.
5. Note the effect of a parallel ac network with both inductive and capacitive elements.

EQUIPMENT REQUIRED

Resistors

1—10-Ω, 1-kΩ

Inductors

1—10-mH

Capacitors

1—0.01-μF

Instruments

1—DMM

1—Oscilloscope

1—Audio oscillator or function generator

1—Frequency counter (if available)

EQUIPMENT ISSUED

TABLE 9.0

Item	Manufacturer and Model No.	Laboratory Serial No.
DMM		
Oscilloscope		
Audio oscillator or function generator		
Frequency counter		

RÉSUMÉ OF THEORY

Kirchhoff's current law as applied to ac circuits states that the phasor sum of the currents entering and leaving a node must equal zero. For example, in a parallel circuit, the source current is the phasor sum of the component (load) currents. For a parallel R-L-C circuit, the magnitude of the source current I_s is given by $I_s = \sqrt{I_R^2 + I_{X_T}^2}$, where $I_{X_T} = I_L - I_C$ (I_L is the current in the inductive branch, I_C is the current in the capacitive branch, and I_R is the current in the resistive branch).

In a parallel R-L-C circuit, it is possible for the magnitude of the source current I_s to be less than a branch current.

Impedances in parallel combine according to the following equation:

$$\frac{1}{Z_T} = \frac{1}{Z_1} + \frac{1}{Z_2} + \frac{1}{Z_3} \tag{9.1}$$

For two impedances in parallel, it is usually more convenient to use the equation in the form

$$Z_T = \frac{Z_1 Z_2}{Z_1 + Z_2} \tag{9.2}$$

In parallel R-L-C circuits, it is possible for the total impedance to be larger than a branch impedance.

To obtain the waveform of a current on an oscilloscope, it is necessary to view the voltage across a resistor. Since the voltage and current of a resistor are related by Ohm's law (a linear relationship), the current and voltage waveforms are always of the same appearance and in phase.

PROCEDURE

Part 1   *R-L* Parallel Network

(a) Construct the network of Fig. 9.1. Insert the measured value of the resistor R. The magnitude of X_L at the applied frequency permits ignoring (on an approximate basis) the effect of the resistor R_l. In other words, assume you have an ideal parallel R-L system.

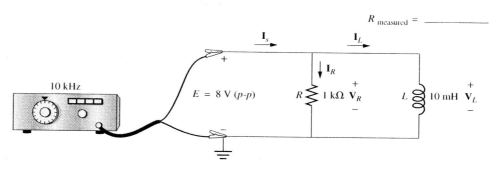

FIG. 9.1

     (b) Using the nameplate inductor value (10 mH) and the measured resistance level for R, calculate the various currents of the network. Note below that the peak-to-peak values are requested, so use peak-to-peak values throughout the calculations. Record the results in Table 9.1.

Calculation:

TABLE 9.1

	Calculated	Measured
$I_{s(p\text{-}p)}$		
$I_{L(p\text{-}p)}$		
$I_{R(p\text{-}p)}$		
$V_{R_{s(p\text{-}p)}}$ (for I_s)		
$V_{R_{s(p\text{-}p)}}$ (for I_L)		

(c) Insert the sensing resistor R_s as shown in Fig. 9.2 to permit a measurement of the source current I_s (the current leaving the generator at the positive terminal will equal that returning to the generator at the negative terminal). Include the measured value of R_s on the diagram in the space provided. Place the scope across the resistor R_s, establishing a common ground with the generator. The resistor R_s is chosen because it is small enough not to affect the general response of the network. Energize the network, set E to 8 V (p-p), measure the peak-to-peak value of the voltage V_{R_s}, and record in the first column of Table 9.1.

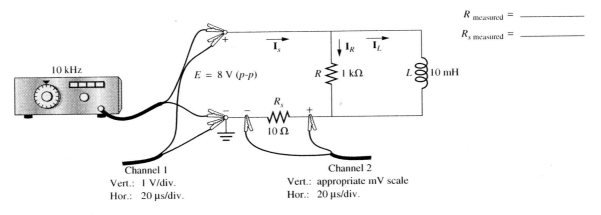

$R_{\text{measured}} =$ ―――――

$R_{s\text{ measured}} =$ ―――――

Channel 1
Vert.: 1 V/div.
Hor.: 20 μs/div.

Channel 2
Vert.: appropriate mV scale
Hor.: 20 μs/div.

FIG. 9.2

(d) Using the measured value of V_{R_s} from part 1(c) and the measured value of the resistor R_s, calculate the peak-to-peak value of the source current I_s and insert in the second column of Table 9.1.

Calculation:

(e) Determine the phase angle θ_s between **E** and **I**$_s$ using the connections of Fig. 9.2. Use the dual-trace method described in Experiment 7. Because $\mathbf{E} = \mathbf{V}_R$ and $\mathbf{I}_R$ is in phase with $\mathbf{V}_R$, this phase angle is also the phase angle between $\mathbf{I}_R$ and $\mathbf{I}_s$. Record D_1 and D_2 in the first row of Table 9.2 and insert the resulting angle in the last column.

Calculation:

(f) Turn off the supply and move the sensing resistor to the position indicated in Fig. 9.3 to permit a measurement of the current I_L. Place the scope across the resistor R_s and be certain to establish a common ground with the generator. Then energize the network, measure the peak-to-peak value of the voltage V_{R_s}, and insert in the first column of Table 9.1.

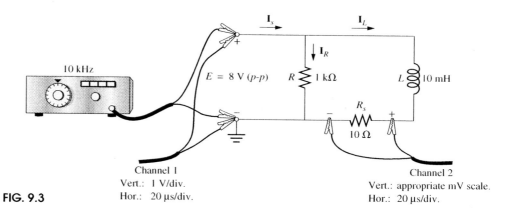

FIG. 9.3

Channel 1
Vert.: 1 V/div.
Hor.: 20 μs/div.

Channel 2
Vert.: appropriate mV scale.
Hor.: 20 μs/div.

(g) Using the measured value of V_{R_s} from part 1(f) and the measured value of the resistor R_s, calculate the peak-to-peak value of the current I_L, and insert in the second column of Table 9.1.

Calculation:

(h) Determine the phase angle θ_L between **E** and $\mathbf{I}_L$ using the connections of Fig. 9.3. Since $\mathbf{E} = \mathbf{V}_R$ and $\mathbf{V}_R$ and $\mathbf{I}_R$ are in phase, the phase angle obtained will also be the phase angle between $\mathbf{I}_R$ and $\mathbf{I}_L$. Record D_1 and D_2 in the second row of Table 9.2 and insert the resulting angle in the last column.

Calculation:

TABLE 9.2

	D_1	D_2	**Angle in Degrees**
θ_s			
θ_L			

(i) Since $V_R = E$, use Ohm's law and the measured value of the resistor R to calculate the peak-to-peak value of the current I_R. Insert the result in the second column of Table 9.1.

Calculation:

(j) Is the source current I_s larger in magnitude than each branch current? Does it have to be? Explain your answers.

(k) Calculate the input impedance using the measured peak-to-peak values of E and I_s. In addition, calculate the reactance of the inductor at the applied frequency. Record the results in Table 9.3.

Calculation:

TABLE 9.3

Z_T	X_L

(l) How does the magnitude of Z_T compare to the magnitude of R or X_L? For a parallel R-L ac network, does Z_T have to be less in magnitude than R or X_L?

(m) Using the input voltage as a reference ($\mathbf{E} = E \angle 0° = 8$ V (p-p) $\angle 0°$) and the measured values of the currents $\mathbf{I}_R$ and $\mathbf{I}_L$, draw a phasor diagram to scale and determine the current $\mathbf{I}_s$. Use all peak-to-peak values.

How does I_s compare to the measured value? From the phasor diagram, determine the angle θ_s between $\mathbf{I}_s$ and $\mathbf{I}_R$ (same as between $\mathbf{E}$ and $\mathbf{I}_s$), the angle θ_L between $\mathbf{I}_s$ and $\mathbf{I}_L$, and the angle θ between $\mathbf{I}_R$ and $\mathbf{I}_L$. Enter the results in Table 9.4.

Comparison:

TABLE 9.4

I_s (diagram)	I_s (measured)	θ_s	θ_L	θ_T

(n) How do the calculated values of θ_s and θ_L compare with the measured values of parts 1(d) and 1(g)? How does the sum $\theta_T = |\theta_s| + |\theta_L|$ for measured values compare to the theoretical value of 90°?

Part 2   *R-C* Parallel Network

(a) Construct the network of Fig. 9.4. Insert the measured value of the resistor *R*.

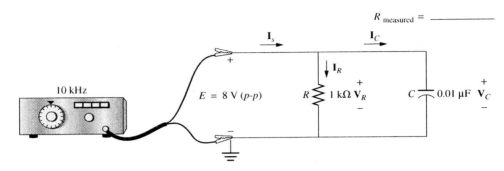

FIG. 9.4

(b) Using the nameplate capacitor value (0.01 μF) and the measured resistance level for *R*, calculate the peak-to-peak values of the currents of the network and record in Table 9.5.

Calculation:

TABLE 9.5

	Calculated	Measured
$I_{S(p\text{-}p)}$		
$I_{C(p\text{-}p)}$		
$I_{R(p\text{-}p)}$		
$V_{R_{S(p\text{-}p)}}$ (for I_S)		
$V_{R_{S(p\text{-}p)}}$ (for I_C)		

(c) Insert the sensing resistor R_s as shown in Fig. 9.5 to permit a measurement of the source current. Include the measured value of R_s on the diagram in the space provided.

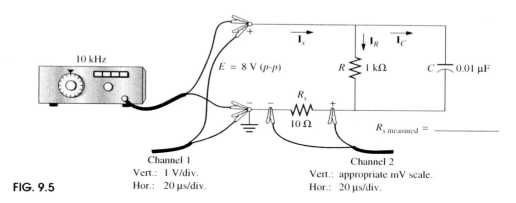

10 kHz

$E = 8$ V $(p\text{-}p)$

I_s I_R I_C

R ⪢ 1 kΩ C ⟂ 0.01 µF

R_s

10 Ω

$R_{s\ \text{measured}}$ = _____

Channel 1
Vert.: 1 V/div.
Hor.: 20 µs/div.

Channel 2
Vert.: appropriate mV scale.
Hor.: 20 µs/div.

FIG. 9.5

Place the scope across the resistor R_s, establishing a common ground with the generator. Energize the network and measure the peak-to-peak value of the voltage V_{R_s} and insert in the first column of Table 9.5.

(d) Using the measured value of V_{R_s} from part 2(c) and the measured value of the resistor R_s, calculate the peak-to-peak value of the source current I_s and insert in the second column of Table 9.5.

Calculation:

(e) Determine the phase angle θ_s between **E** and **I**$_s$ using the connections of Fig. 9.5. Because **E** = **V**$_R$ and **I**$_R$ is in phase with **V**$_R$, it is also the phase angle between **I**$_R$ and **I**$_s$. Record D_1 and D_2 in Table 9.6 and insert the resulting angle in the last column.

Calculation:

TABLE 9.6

	D_1	D_2	Angle in Degrees
θ_s			
θ_C			

(**f**) Turn off the supply and move the sensing resistor to the position indicated in Fig. 9.6 to permit a measurement of the current I_C. Place the scope across the resistor R_s and be certain to establish a common ground with the generator. Then energize the network, measure the peak-to-peak value of the voltage V_{R_s}, and insert in the first column of Table 9.5.

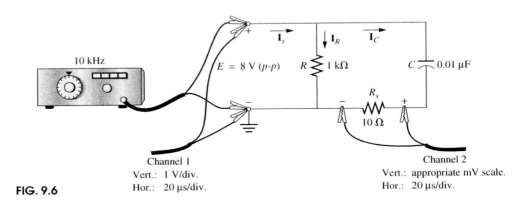

FIG. 9.6

Channel 1
Vert.: 1 V/div.
Hor.: 20 µs/div.

Channel 2
Vert.: appropriate mV scale.
Hor.: 20 µs/div.

(**g**) Using the measured value of V_{R_s} from part 2(f) and the measured value of the resistor R_s, calculate the peak-to-peak value of the current I_C and insert in the second column of Table 9.5.

Calculation:

(**h**) Determine the phase angle θ_C between **E** and **I**$_C$ using the connections of Fig. 9.6. Because $\mathbf{E} = \mathbf{V}_R$ and $\mathbf{V}_R$ and $\mathbf{I}_R$ are in phase, the phase angle obtained will also be the phase angle between $\mathbf{I}_R$ and $\mathbf{I}_C$. Record D_1 and D_2 in the second row of Table 9.6 and insert the resulting angle in the last column.

Calculation:

(**i**) Since $V_R = E$, use Ohm's law and the measured value of the resistor R to calculate the peak-to-peak value of the current I_R. Insert the result in the second column of Table 9.5.

Calculation:

(**j**) Is the source current I_s larger in magnitude than each branch current? Does it have to be?

(**k**) Calculate the input impedance using the peak-to-peak values of E and I_s. In addition, calculate the reactance of the capacitor at the applied frequency. Record the results in Table 9.7.

TABLE 9.7

Z_T	X_L

(**l**) How does the magnitude of Z_T compare to the magnitude of R or X_C? For a parallel R-C network, does Z_T have to be less in magnitude than R or X_C?

(**m**) Using the input voltage as a reference ($\mathbf{E} = E\angle0° = 8$ V $(p\text{-}p)\angle0°$) and the measured values of the currents $\mathbf{I}_R$ and $\mathbf{I}_C$, draw a phasor diagram to scale and determine the current $\mathbf{I}_s$. Use all peak-to-peak values.

How does I_s compare to the measured value? From the phasor diagram, determine the angle θ_s between $\mathbf{I}_s$ and $\mathbf{I}_R$ (same as between $\mathbf{E}$ and $\mathbf{I}_s$), the angle θ between $\mathbf{I}_s$ and $\mathbf{I}_C$, and the angle θ_1 between $\mathbf{I}_R$ and $\mathbf{I}_C$. Enter the results in Table 9.8.

Comparison:

TABLE 9.8

I_s (diagram)	I_s (calculated)	θ_s	θ_C	θ_T

(n) How do the calculated values of θ_s and θ_C compare with the measured values of parts 2(d) and 2(g)? How does the sum $\theta_T = |\theta_s| + |\theta_C|$ for measured values compare to the theoretical value of $90°$?

Part 3 R-L-C Parallel Network

(a) Construct the network of Fig. 9.7. Insert the measured value of the resistor R.

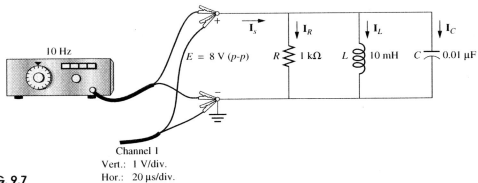

10 Hz

$E = 8$ V $(p\text{-}p)$ $R \gtrless 1$ kΩ L 10 mH $C \top 0.01$ µF

Channel 1
Vert.: 1 V/div.
Hor.: 20 µs/div.

FIG. 9.7

(b) Using nameplate values for the inductance and capacitance and the measured resistor value, calculate the peak-to-peak values of the currents of the network and record in Table 9.9.

Calculation:

TABLE 9.9

	Calculated	Measured
$I_{S(p\text{-}p)}$		
$I_{L(p\text{-}p)}$		
$I_{C(p\text{-}p)}$		
$I_{R(p\text{-}p)}$		
$V_{R_{S(p\text{-}p)}}$ (for I_S)		
$V_{R_{S(p\text{-}p)}}$ (for I_L)		
$V_{R_{S(p\text{-}p)}}$ (for I_C)		

(c) Using a sensing resistor as described in the earlier parts of this experiment, determine the measured value of the peak-to-peak value of the source current. Record the peak-to-peak voltage reading and the resulting current in the second column of Table 9.9. Compare with the calculated value of part 3(b). Show all work below.

Calculation:

(d) Using a sensing resistor, determine the peak-to-peak value of the currents I_L and I_C. Record the peak-to-peak voltage readings and the resulting currents in the second column of Table 9.9. Compare with the calculated values of part 3(b). Show all work below. In this case, the positions of L and C will have to be interchanged to be sure the sensing resistor is measuring the current in only one branch of the network.

Calculation:

(e) Calculate the peak-to-peak value of the current I_R using the fact that $V_R = E$. Insert the results in the second column of Table 9.9.

Calculation:

(f) Using the input voltage as a reference ($\mathbf{E} = E \angle 0° = 8$ V $(p\text{-}p)\angle 0°$) and the measured values of all the currents, draw a phasor diagram and determine the phase angle θ_s between the applied voltage and the input current. Record the result in the first column of Table 9.10.

TABLE 9.10

	Method One	Method Two
θ_s		
D_1		
D_2		
Z_T		

(g) Measure the phase angle between the applied voltage and the source current using a technique described earlier in this laboratory experiment and compare with the calculated value of part 3(f). Record D_1, D_2, and the resulting angle in Table 9.10.

Calculation:

(h) Using the calculated value of I_s from part 3(b) and the applied voltage, calculate the magnitude of the total impedance of the network and record in the first column of Table 9.10. How does it compare to the value of part 3(i)?

Calculation:

Comparison:

(i) Using the measured value of I_s from part 3(c) and the applied voltage, calculate the magnitude of the total impedance of the network and record in the second column of Table 9.10.

Calculation:

EXERCISES

1. Determine X_L, X_C, and then Z_T for the network of Fig. 9.7 at a frequency of 10 kHz.

$X_L = $ _____

$X_C = $ _____

$Z_T = $ _____

2. For the case of Problem 1, is Z_T less than each of the parallel components at 10 kHz?

3. For a parallel R-L-C ac network, are there frequencies where the total impedance can be greater than, less than, or equal to the smallest parallel impedance (R, X_L, or X_C) of the network? Don't go into a lot of detail. Just discuss the subject in general and the effect of frequency on the possibilities.

4. Using PSpice or Multisim, determine all the currents for the network of Fig. 9.7 and compare with the results from the measured column of Table 9.9. Attach all appropriate printouts.

Series-Parallel Sinusoidal Circuits

OBJECTIVES

1. Measure the currents of a series-parallel R-L and R-C network using sensing resistors.
2. Demonstrate the Pythagorean relationship between the currents of the networks.
3. Experimentally measure the phase angles associated with the currents of the network.
4. Calculate the input impedance of a parallel network using measured values.

EQUIPMENT REQUIRED

Resistors

1—10-Ω, 470-Ω, 1-kΩ (1/4-W)

Inductors

1—10-mH

Capacitors

1—0.02-μF

Instruments

1—DMM

1—Oscilloscope

1—Audio oscillator or function generator

1—Frequency counter (if available)

EQUIPMENT ISSUED

TABLE 10.0

Item	Manufacturer and Model No.	Laboratory Serial No.
DMM		
Oscilloscope		
Audio oscillator or function generator		
Frequency counter		

RÉSUMÉ OF THEORY

In the previous experiments, we showed that Kirchhoff's voltage and current laws hold for ac series and parallel circuits. In fact, all the previously used rules, laws, and methods of analysis apply equally well for both dc and ac networks. The major difference for ac circuits is that we now use reactance, resistance, and impedance instead of solely resistance.

Consider the common series-parallel circuit of Fig. 10.1.

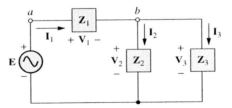

FIG. 10.1

The various impedances shown could be made up of many elements in a variety of configurations. No matter how varied or numerous the elements might be, $\mathbf{Z}_1$, $\mathbf{Z}_2$, and $\mathbf{Z}_3$ represent the total impedance for that branch. For example, $\mathbf{Z}_1$ might be as shown in Fig. 10.2.

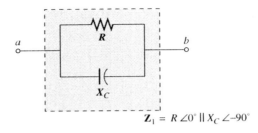

$$\mathbf{Z}_1 = R \angle 0° \parallel X_C \angle -90°$$

FIG. 10.2

The currents and voltages in Fig. 10.1 can be found by applying any of the methods outlined for dc networks; in particular,

$$\mathbf{Z}_T = \mathbf{Z}_1 + \mathbf{Z}_2 \parallel \mathbf{Z}_3 \qquad \mathbf{I}_s = \mathbf{I}_1 = \frac{\mathbf{E}}{\mathbf{Z}_T}$$

and

$$\mathbf{I_2} = \frac{\mathbf{Z_3} \, \mathbf{I_s}}{\mathbf{Z_3} + \mathbf{Z_2}} \qquad \text{(Current Divider Rule)}$$

$\mathbf{I_3}$ can be determined using Kirchhoff's current law:

$$\mathbf{I_3} = \mathbf{I_s} - \mathbf{I_2}$$

Ohm's law provides the voltages:

$$\mathbf{V_1} = \mathbf{I_1 Z_1} \qquad \mathbf{V_2} = \mathbf{I_2 Z_2} \qquad \mathbf{V_3} = \mathbf{I_3 Z_3}$$

Keep in mind that the voltages and currents are phasor quantities that have both magnitude and an associated angle.

PROCEDURE

Part 1   *R-L* Series-Parallel Network

(a) Construct the network of Fig. 10.3. Insert the measured value for each resistor. At the applied frequency, the resistor R_l can be ignored in comparison with the other elements of the network. Set the source to 8 V (*p-p*) with the oscilloscope.

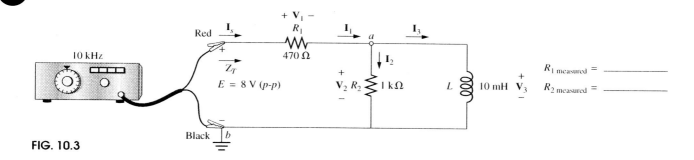

FIG. 10.3

(b) Calculate the magnitude of the input impedance Z_T of the network at $f = 10$ kHz. Use measured resistor values and ignore the effects of R_l (the internal resistance of the coil). Record the result in Table 10.1.

Calculation:

TABLE 10.1

	Calculated	Measured
Z_T		
$I_{s(p\text{-}p)}$		
$V_{1(p\text{-}p)}$		
$V_{2(p\text{-}p)}$		
$V_{3(p\text{-}p)}$		
$I_{2(p\text{-}p)}$		
$I_{3(p\text{-}p)}$		

   (c) Using the results of part 1(b) and Ohm's law, calculate the peak-to-peak value of the source current $I_s = I_1$ and record in the calculated column of Table 10.1.

Calculation:

   (d) Reverse the leads to the generator (to insure a common ground between the generator and oscilloscope) and measure the peak-to-peak value of the voltage V_1 with the oscilloscope. Record in the measured column of Table 10.1.
   Using the measured value for the voltage, calculate the peak-to-peak value of the current $I_s = I_1$. Record the result in the measured column of Table 10.1.

Calculation:

   How does the magnitude of $I_{s(p\text{-}p)}$ compare with the results of part 1(c)?

(e) Determine $\mathbf{Z}_T$ from the measured value of $\mathbf{I}_{s(p\text{-}p)}$ and the source voltage using the following equation:

$$Z_T = \frac{E_{(p\text{-}p)}}{I_{s(p\text{-}p)}}$$

Insert the result in the measured column of Table 10.1.

Calculation:

How does the measured value of Z_T compare with the calculated level of part 1(b)?

(f) Calculate the peak-to-peak values of the voltage $\mathbf{V}_1$, $\mathbf{V}_2$, and $\mathbf{V}_3$ and insert in the calculated column of Table 10.1.

Calculation:

(g) **Reestablish the network of Fig. 10.3** with R_1 connected as shown and measure the peak-to-peak values of $\mathbf{V}_2$ and $\mathbf{V}_3$ with the oscilloscope by connecting the ground terminal of the scope channel to point b and the high end of the channel to point a. Record the readings in the measured column of Table 10.1.

What is their relationship? Why?

(h) Using the results of part 1(f), calculate the peak-to-peak values of I_1, I_2, and I_3 and insert in the calculated column of Table 10.1.

Calculation:

(i) Using the result of part 1(g), calculate the peak-to-peak values of the current I_2 and I_3 and record in the measured column of Table 10.1.

Calculation:

How do the measured values compare with the calculated levels of part 1(h)?

(j) Using the results of parts 1(d) and 1(i), determine whether the following relationship is satisfied using peak-to-peak values:

$$I_1 = I_s = \sqrt{I_2^2 + I_3^2}$$

(k) Devise a method to measure the phase angle between E and V_2 ($= V_3$). Check your method with the instructor and then determine the angle. **Be sure to note which leads or lags.** Record the results in the top row of Table 10.2.

Calculation:

TABLE 10.2

	D_1	D_2	θ
E and **V**$_2$			
E and **I**$_s$			
E and **V**$_1$			

(**l**) Devise a method to measure the phase angle between **E** and **I**$_s$ = **I**$_1$. Check your method with the instructor and then determine the angle. Be sure to note which leads or lags. Record the results in the second row of Table 10.2.

Calculation:

(**m**) Using the results from the bottom row of Table 10.2, calculate the phase angle between **E** and **V**$_1$. Record the result in Table 10.2.

Calculation:

(**n**) Using **E** = 8 V$(p$-$p)\angle 0°$, draw the phasor diagram of **E**, **V**$_1$, **V**$_2$, and **I**$_s$ using the above results with the peak-to-peak values.

Is the vector sum of $\mathbf{V}_1$ and $\mathbf{V}_2$ equal to $\mathbf{E}$, as required by Kirchhoff's voltage law? Do not perform any mathematical calculations but simply note whether the sum of the vectors $\mathbf{V}_1$ and $\mathbf{V}_2$ on the diagram would result in the source voltage $\mathbf{E}$.

Part 2   R-C Series-Parallel Network

(a) Construct the network of Fig. 10.4. Insert the measured resistor values. Set the source to 8 V (p-p) with the oscilloscope.

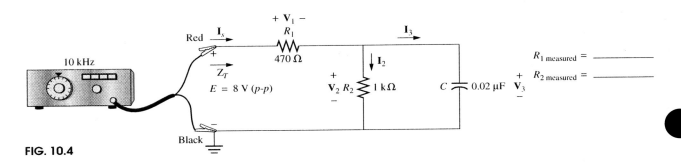

FIG. 10.4

(b) Calculate the magnitude of the input impedance Z_T at $f = 10$ kHz and record the result in Table 10.3.

TABLE 10.3

	Calculated	Measured
Z_T		
$I_{s(p-p)}$		
$V_{1(p-p)}$		
$V_{2(p-p)}$		
$V_{3(p-p)}$		
$I_{2(p-p)}$		
$I_{3(p-p)}$		

     (c) Using the result of part 1(b) and Ohm's law, calculate the peak-to-peak value of the source current $I_s = I_1$ and record in the calculated column of Table 10.3.

Calculation:

     (d) Reverse the leads to the generator (to ensure a common ground between the generator and oscilloscope) and measure the peak-to-peak value of the voltage V_1 with the oscilloscope. Record the result in the measured column of Table 10.3.

     Using the measured value for the voltage, calculate the peak-to-peak value of the current $I_s = I_1$. Record the result in the measured column of Table 10.3.

Calculation:

How does the magnitude of the measured $I_{s(p-p)}$ compare to the results of part 1(c)?

(e) Determine Z_T from the measured value of $I_{s(p-p)}$ and the source voltage using the following equation:

$$Z_T = \frac{E_{(p-p)}}{I_{s(p-p)}}$$

Insert the result in the measured column of Table 10.3.

Calculation:

How does the measured value of Z_T compare with the calculated level of part 1(b)?

(f) Calculate the peak-to-peak values of the voltage V_1, V_2, and V_3 and insert in the calculated column of Table 10.3.

Calculation:

(g) **Reestablish the network of Fig. 10.4** with R_1 connected as shown and measure the peak-to-peak values of V_2 and V_3 with the oscilloscope by connecting the ground terminal of the scope channel to point b and the high end of the channel to point a. Record the readings in the measured column of Table 10.3.

What is their relationship? Why?

(h) Using the result of part 1(f), calculate the peak-to-peak values of the current I_1, I_2, and I_3, and insert in the calculated column of Table 10.3.

Calculation:

(i) Using the results of part 1(g), calculate the peak-to-peak values of the current I_2 and I_3 and record in the measured column of Table 10.3.

Calculation:

How do the measured values compare with the calculated levels of part 1(h)?

(j) Using the results of parts 1(d) and 1(i), determine whether the following relationship is satisfied using peak-to-peak values:

$$I_1 = I_s = \sqrt{I_2^2 + I_3^2}$$

(k) Devise a method to measure the phase angle between **E** and I_s. Check the method with your instructor and then determine the angle. Be sure to note which leads or lags. Record the result in the top row of Table 10.4.

Calculation:

TABLE 10.4

	D_1	D_2	θ
E and $\mathbf{I}_s$			
E and $\mathbf{I}_2$			
$\mathbf{I}_2$ and $\mathbf{I}_3$			

(**l**) Devise a method to measure the phase angle between **E** and $\mathbf{I}_2$. Check the method with your instructor and then determine the angle. Be sure to note which leads or lags. Record the result in the second row of Table 10.4.

Calculation:

(**m**) Devise a method to determine the phase angle between $\mathbf{I}_2$ and $\mathbf{I}_3$. Check the method with your instructor and then determine the angle. Record the result in the last row of Table 10.4.

Calculation

(**n**) If $\mathbf{E} = E \angle 0°$, write $\mathbf{V}_1$ and $\mathbf{V}_2$ in phasor form using the results obtained above. Use peak-to-peak values for the magnitudes.

$$V_1 = \underline{\hspace{3cm}}$$

$$V_2 = \underline{\hspace{3cm}}$$

(o) Is the following Kirchhoff's voltage law relationship satisfied? Use peak-to-peak values.

$$E = \sqrt{V_1^2 + V_2^2}$$

If not, why not?

Part 3   *R-L-C* Series-Parallel Network

(a) Construct the network of Fig. 10.5. Insert the measured resistor values.

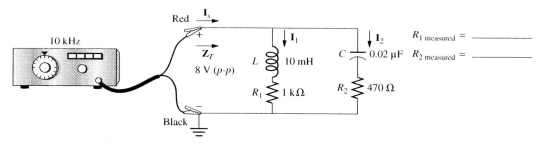

FIG. 10.5

(b) Calculate Z_T at a frequency of 10 kHz using the nameplate values of the elements ($R_1 = 1$ kΩ, $R_2 = 470$ Ω, $L = 10$ mH, $C = 0.02$ μF) and record in Table 10.5.

TABLE 10.5

	Calculated	Measured
Z_T		
$I_{s(p\text{-}p)}$		
$I_{1(p\text{-}p)}$		
$I_{2(p\text{-}p)}$		

(c) Calculate the peak-to-peak value of I_s with $E = 8$ V (p-p). Record the result in Table 10.5.

Calculation:

(d) Calculate the peak-to-peak levels of I_1 and I_2 and record the results in Table 10.5.

Calculation:

(e) Energize the network and measure the voltages $\mathbf{V}_{R_1}$ and $\mathbf{V}_{R_2}$ with the oscilloscope. Then calculate the peak-to-peak values of the currents I_1 and I_2 and insert the results in the measured column of Table 10.5.

Calculation:

(f) Using the results of part 3(e), calculate the peak-to-peak value of I_s and insert in the measured column of Table 10.5.

Calculation:

How does it compare to the calculated value of part 3(c)?

(g) Using the result of part 3(f), calculate the input impedance and insert as the measured value in Table 10.5.

Calculation:

How does it compare to the calculated value of part 3(b)?

(h) Use the oscilloscope to determine the phase angle between **E** and **I**$_1$ and record the results in the first row of Table 10.6.

Calculation:

TABLE 10.6

	D_1	D_2	θ
E and **I**$_1$			
E and **I**$_2$			

(i) Use the oscilloscope to determine the phase angle between $\mathbf{E}$ and $\mathbf{I}_2$ and record the results in the second row of Table 10.6.

Calculation:

(j) With $\mathbf{E} = E \angle 0°$, sketch the phasor diagram of the currents and source voltage of the network—that is, a phasor diagram including $\mathbf{I}_1$, $\mathbf{I}_2$, $\mathbf{I}_s$, and $\mathbf{E}$. The magnitude and location of $\mathbf{I}_s$ can be derived from the vectors $\mathbf{I}_1$ and $\mathbf{I}_2$ and an application of Kirchhoff's current law.

(k) Determine the magnitude of the current $\mathbf{I}_s$ from the phasor diagram of part 3(j). How does it compare to the magnitude determined in part 3(c)? Complete Table 10.7.

TABLE 10.7

$I_{s_{(p\text{-}p)}}$(diagram)	
$I_{s_{(p\text{-}p)}}$(part 3(c))	

(l) Devise an experimental technique to measure the phase angle between I_s and I_1. Sketch the hookup in the space below, showing all the connections to the oscilloscope. If a sensing resistor is employed, what standard value would be appropriate? Record all the results in Table 10.8.

Calculation:

TABLE 10.8

D_1	D_2	θ	R

EXERCISES

1. Determine the phase angle between V_1 and V_2 for the network of Fig. 10.3 and compare to the measured value of part 1(n).

θ_s (calculated) = _____, θ_s (part 1(n)) = _____

2. Determine the phase angle between $I_1 = I_s$ and I_2 for the network of Fig. 10.4 and draw a complete phasor diagram for all the currents of Fig. 10.4 using the results of part 2.

$\theta_s =$ _____

3. Calculate the phase angle between I_1 and I_2 for the network of Fig. 10.5 and compare to the measured value of part 3(j).

θ_1 (calculated) = _____, θ_1 (part 3(j)) = _____

Thevenin's Theorem and Maximum Power Transfer

OBJECTIVES

1. Verify Thevenin's theorem through experimental measurements.
2. Become acutely aware of the differences between applying Thevenin's theorem to an ac network as compared to a dc system.
3. Demonstrate the validity of the conditions for maximum power transfer in an ac network.

EQUIPMENT REQUIRED

Resistors

1—10-Ω, 470-Ω, 1-kΩ, 1.2-kΩ, 2.2-kΩ, 3.3-kΩ, 6.8-kΩ (1/4-W)
1—0–1-kΩ Potentiometer

Capacitors

1—0.0047-μF, 0.01-μF, 0.02-μF, 0.047-μF, 0.1-μF, 1-μF
2—0.02-μF

Inductors

1—10-mH

Instruments

1—DMM
1—Oscilloscope
1—Audio oscillator or function generator
1—Frequency counter (if available)

EQUIPMENT ISSUED

TABLE 11.0

Item	Manufacturer and Model No.	Laboratory Serial No.
DMM		
Oscilloscope		
Audio oscillator or function generator		
Frequency counter		

RÉSUMÉ OF THEORY

Thevenin's theorem states that any two-terminal linear ac network can be replaced by an equivalent circuit consisting of a voltage source in series with an impedance. To apply this theorem, follow these simple steps:

1. Remove the portion of the network across which the Thevenin equivalent circuit is found.
2. Replace all voltage sources by a short-circuit equivalent and all current sources by an open-circuit equivalent.
3. Calculate $\mathbf{Z}_{Th}$ across the two terminals in question.
4. Replace all sources.
5. Calculate $\mathbf{E}_{Th}$, which is the voltage across the terminals in question.
6. Draw the Thevenin equivalent circuit and replace the portion of the circuit originally removed.
7. Solve for the voltage or current originally desired.

EXAMPLE   Using Thevenin's theorem, find the current through R in Fig. 11.1.

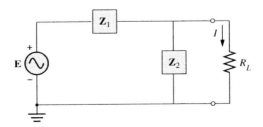

FIG. 11.1

Solution

Step 1:	Remove R. See Fig. 11.2(a).
Step 2:	Replace **E** by a short-circuit equivalent. See Fig. 11.2(a).
Step 3:	Solve for $\mathbf{Z}_{Th}$. In this case, $\mathbf{Z}_{Th} = \mathbf{Z}_1 \| \mathbf{Z}_2$. See Fig. 11.2(a).
Step 4:	Replace **E**. See Fig. 11.2(b).
Step 5:	Calculate $\mathbf{E}_{Th}$ from

$$\mathbf{E}_{Th} = \frac{\mathbf{Z}_2}{\mathbf{Z}_1 + \mathbf{Z}_2}\mathbf{E}$$

See Fig. 11.2(b).

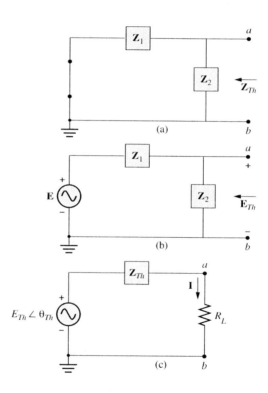

FIG. 11.2

Step 6:	Draw the equivalent circuit and replace R. See Fig. 11.2(c).
Step 7:	Calculate **I** from

$$\mathbf{I} = \frac{\mathbf{E}_{Th}}{\mathbf{Z}_{Th} + \mathbf{Z}_{R_L}}$$

MAXIMUM POWER TRANSFER THEOREM

The maximum power transfer theorem states that for circuits with ac sources, maximum power will be transferred to a load when the load impedance is the conjugate of the Thevenin impedance across its terminals. See Fig. 11.3.

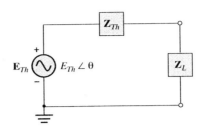

FIG. 11.3

For maximum power transfer, if

$$\mathbf{Z}_{Th} = |\mathbf{Z}_{Th}| \angle \theta$$

then

$$\mathbf{Z}_L = |\mathbf{Z}_{Th}| \angle -\theta$$

PROCEDURE

Part 1

(a) Construct the network of Fig. 11.4. Insert the measured values of R_L, R_1, and R_2 in Fig. 11.4.

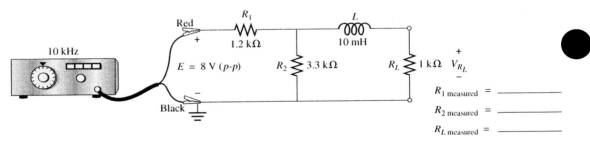

FIG. 11.4

(b) With E set at 8 V (p-p), measure $V_{R_{L(p-p)}}$ with the oscilloscope and record in the first column of Table 11.1.

TABLE 11.1

	Original	Thevenin Equivalent
$V_{R_{L(p-p)}}$ $(R_L = 1\text{k}\Omega)$		
$\mathbf{E}_{Th}$		
$\mathbf{Z}_{Th}$		
$V_{R_{L(p-p)}}$ $(R_L = 6.8\text{k}\Omega)$		

(c) Remove the resistor R_L and measure the open-circuit voltage across the resulting terminals with the oscilloscope. This is the magnitude of $\mathbf{E}_{Th}$. Record this peak-to-peak value in the first column of Table 11.1.

(d) Using the measured resistance levels, calculate the magnitude of $\mathbf{E}_{Th}$ and its associated angle. Assume that $\mathbf{E} = E \angle 0° = 8\ \text{V}(p\text{-}p) \angle 0°$. Show all work! Organize your presentation. Record both the magnitude (peak-to-peak) and angle in the second column of Table 11.1.

Calculation:

How does the magnitude of the calculated value compare with the measured value of part 1(c)?

(e) Calculate $\mathbf{Z}_{Th}$ (magnitude and angle) using the measured resistor values and the nameplate inductor level of 10 mH. The applied frequency is 10 kHz. Determine the resistive and reactive components of $\mathbf{Z}_{Th}$ by converting $\mathbf{Z}_{Th}$ to rectangular form. Record $\mathbf{Z}_{Th}$ in the rectangular form in the first column of Table 11.2.

Calculation:

(f) Insert the measured value of $E_{Th(p\text{-}p)}$ from part 1(c) in Fig. 11.5. Set the 1-kΩ potentiometer R of Fig. 11.5 to the value calculated in part 1(e). Use 10 mH for the inductor, since X_L of part 1(e) is defined solely by the inductor of Fig. 11.4. The Thevenin equivalent circuit for the network to the left of the resistor R_L of Fig. 11.4 has now been established in Fig. 11.5. Turn on the supply, set to the indicated value, and measure the peak-to-peak level of the voltage across the resistor R_L with the oscilloscope. Record the peak-to-peak reading in the second column of Table 11.1.

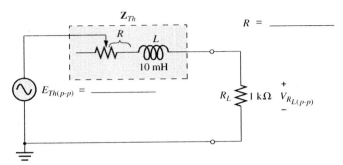

FIG. 11.5

How does this value compare with the measured level of part 1(b) for the original network? Has the equivalence of the Thevenin circuit been verified?

(g) Change R_L in Fig. 11.5 to a 6.8-kΩ resistor and using the Thevenin equivalent network calculate the resulting voltage across the new resistance level. Record the peak-to-peak value in the second column of Table 11.1.

Calculation:

Change R_L in Fig. 11.4 to 6.8 kΩ and measure the resulting voltage across the new resistance level using the oscilloscope. Record the resulting peak-to-peak reading in the first column of Table 11.1.

How do the two levels of $V_{R_{L(p-p)}}$ compare?

Do the preceding measurements verify the fact that the Thevenin equivalent circuit is valid for any level of R_L?

Part 2   Determining Z_{Th} Experimentally

The Thevenin impedance can be determined experimentally following the procedure outlined in the Résumé of Theory. The process of replacing the source by a short-circuit equivalent is the same as removing the source and reconstructing the network, as shown in Fig. 11.6. By applying a source to the load terminals with a sensing resistor, the magnitude of the Thevenin impedance can be determined from

$$|Z_{Th}| = \frac{E_{p\text{-}p}}{I_{p\text{-}p}} \tag{11.1}$$

with

$$I_{p\text{-}p} = \frac{V_{R_{s(p\text{-}p)}}}{R_{s(\text{measured})}}$$

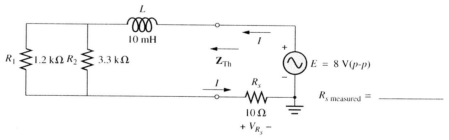

FIG. 11.6

The angle associated with the Thevenin impedance can be determined by finding the phase angle between **E** and **I** (same phase angle as $\mathbf{V}_{R_s}$) using the oscilloscope.

(a) Construct the network of Fig. 11.6. Insert the measured value of R_s.

(b) Set E to 8 V (p-p) with the oscilloscope and measure the voltage V_{R_s} with the oscilloscope. Then calculate the peak-to-peak value of the current I using Ohm's law. Record both results in Table 11.2.

Calculation:

TABLE 11.2

$V_{R_s(p\text{-}p)}$	
$I_{(p\text{-}p)}$	
Z_{Th}	
D_1	
D_2	
θ	

Determine the magnitude of Z_{Th} using Eq. (11.1) and ignoring the effects of R_s. Record the result in Table 11.2.

Calculation:

(c) Determine the phase angle between **E** and V_{R_s} (same as **I**) using the oscilloscope. Show all calculations. The angle between **E** and **I** is the same one associated with the Thevenin impedance. Record the measurements and calculated angle in Table 11.2.

Calculation:

(d) Using the results of parts 2(b) and 2(c), write the Thevenin impedance in polar form and then convert to the rectangular form to define the levels of R and X_L. Record the result in rectangular form in the second column of Table 11.1.

(e) How do the rectangular forms of Z_{Th} in Table 11.1 compare?

Part 3 Maximum Power Transfer

(a) Construct the circuit of Fig. 11.7. Include the measured resistor values.

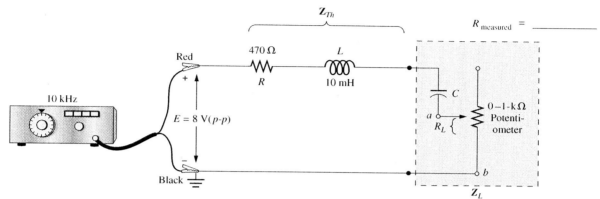

FIG. 11.7

(b) Set the 0–1-kΩ potentiometer to $R_L = R_{\text{measured}}$, as required for maximum power transfer. Place this value in each row of Table 11.3. In addition, record R_L in Table 11.4.

TABLE 11.3

R_L	X_L	$|Z_{Th}|$	C^*	X_C	$|Z_L|$	$V_{ab(p\text{-}p)}$	$P_L = (V_{ab(p\text{-}p)})^2/8R_L$
			0.0047 μF				
			0.01 μF				
			0.0147 μF				
			0.0247 μF				
			0.047 μF				
			0.1 μF				
			1 μF				

C^*: some capacitance levels are the result of placing the given capacitors in parallel.

(c) For $f = 10$ kHz, calculate X_L and insert (the same value) in each row of Table 11.3.

(d) Using the results of parts 2(b) and 2(c), calculate the magnitude of the Thevenin impedance Z_{Th} and insert (the same value) in each row of Table 11.3.

(e) For each value of C, calculate X_C at $f = 10$ kHz and insert in Table 11.3.

(f) Calculate the magnitude of $\mathbf{Z}_L$ with $R_L = R_{\text{measured}}$ and insert in each row of Table 11.3.

(g) Energize the network and, for each capacitance value, measure the peak-to-peak value of V_{ab} with the oscilloscope and insert in Table 11.3. (Note that 0.0247 μF = 0.02 μF + 0.0047 μF.) To save time, use the dual-trace capability of your oscilloscope to measure $V_{ab(p\text{-}p)}$ and maintain E at 8 V $(p\text{-}p)$.

(h) Calculate the power delivered to the load using $P_L = (V_{ab(p-p)})^2/8R_L$ derived from

$$P_L = \frac{V_{ab(\text{rms})}^2}{R_L} = \frac{\left(\frac{1}{\sqrt{2}}\left(\frac{V_{ab_{(p-p)}}}{2}\right)\right)^2}{R_L} = \frac{\left(\frac{V_{ab_{(p-p)}}^2}{8}\right)}{R_L} = \frac{(V_{ab_{(p-p)}}^2)}{8R_L} \quad \textbf{(11.2)}$$

(i) Plot P_L versus X_C on Graph 11.1. Finish off the plot as well as you can with the data you have.

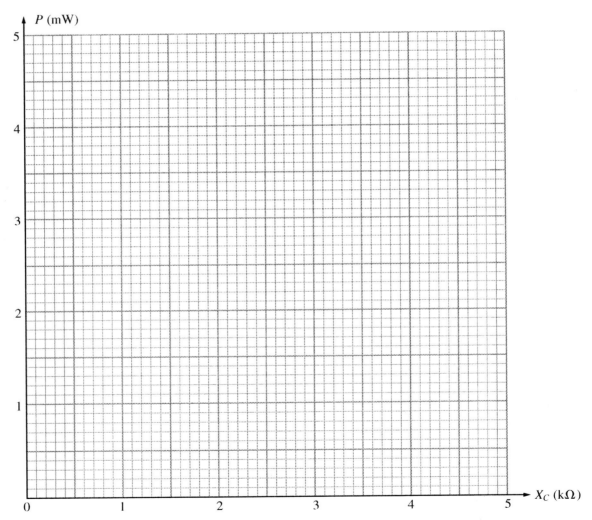

GRAPH 11.1

(j) Using Graph 11.1, determine X_C when the power to the load is a maximum. How does it compare to the level of X_L of the second column of Table 11.3? Record the level of X_C in Table 11.4.

TABLE 11.4

R_L	
X_C	
C	

(k) Maximum power transfer requires that $X_C = X_L$. Has this condition been verified by the experimental data?

The theory also requires that $|Z_L| = |Z_{Th}|$ for maximum power transfer to the load. Has this condition also been met when P_L is a maximum?

(I) Assuming that L is exactly, 10 mH, what value of capacitance will ensure that $X_C = X_L$ at $f = 10$ kHz? How close do we come to this value in Table 11.3? Record the value of C in Table 11.4.

Part 4   Maximum Power Transfer (The Resistive Component)

(a) Reconstruct the network of Fig. 11.7 with C set to the value that resulted in maximum power transfer in part 2. Record the value of C to be used in Table 11.5.

(b) In this part of the experiment, we will determine whether maximum power to the load is, in fact, delivered when $R_L = R_{Th}$ by varying the load resistance. Set R_L to the values listed in Table 11.6 and measure the voltage $V_{ab_{(p-p)}}$ with the oscilloscope. For each row of Table 11.6, insert $V_{ab_{(p-p)}}$ and calculate the power to the load. Record the measured value of $R_{Th} = R$ in Table 11.5.

TABLE 11.5

C	
R_{Th}	
$R_L(\text{for } P_{max})$	

      (c) Plot P_L versus R_L on Graph 11.2, clearly indicating each plot point and labeling the curve. Finish off the curve as well as you can with the data you have.

      (d) At what value (using Graph 11.2) of R_L is maximum power delivered to the load? Record in Table 11.5. How does this value compare to the resistance R_{Th}? Should they be close in value?

TABLE 11.6

R_L	$V_{ab_{(p-p)}}$	$P_L = V_{ab_{(p-p)}}{}^2/8R_L$
100 Ω		
300 Ω		
400 Ω		
500 Ω		
600 Ω		
800 Ω		
1000 Ω		
$R_L = R_{Th} = $ _____		

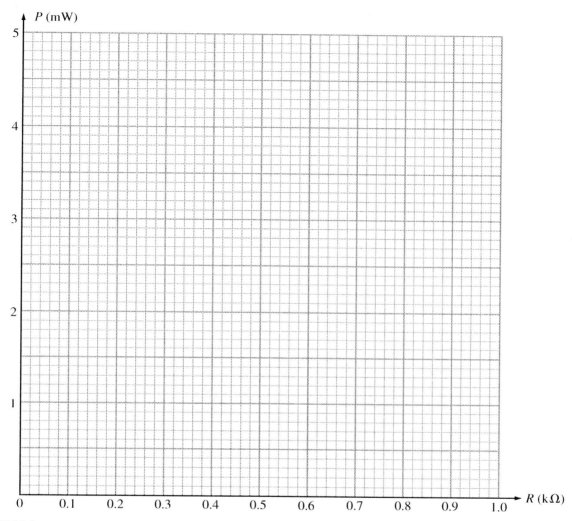

GRAPH 11.2

EXERCISE

1. Using PSpice or Multisim, plot the power to the load R_L in Fig. 11.7 vs. values of R_L from 100 Ω to 1000 Ω using $C = 25$ nF. Compare the resulting plot to that obtained in Graph 11.2. Attach all pertinent printouts.

Series Resonant Circuits

OBJECTIVES

1. Validate the basic equation for the resonant frequency of a series resonant circuit.
2. Plot the various voltages and current for a series resonant circuit versus frequency.
3. Verify that the input impedance is a minimum at the resonant frequency.
4. Demonstrate the relationship between the Q of a series resonant circuit and the resulting bandwidth.

EQUIPMENT REQUIRED

Resistors

1—33-Ω, 220-Ω

Inductors

1—10-mH

Capacitors

1—0.1-μF, 1-μF

Instruments

1—DMM

1—Audio oscillator or function generator

1—Frequency counter (strongly recommended)

EQUIPMENT ISSUED

TABLE 12.0

Item	Manufacturer and Model No.	Laboratory Serial No.
DMM		
Audio oscillator or function generator		
Frequency counter		

RÉSUMÉ OF THEORY

In a series R-L-C circuit, there exists one frequency at which $X_L = X_C$ or $\omega L = 1/\omega C$. At this frequency, the circuit is in resonance and the input voltage and current are in phase. At resonance, the circuit is resistive in nature and has a minimum value of impedance or a maximum value of current.

The resonant radian frequency $\omega_s = 1/\sqrt{LC}$ and frequency $f_s = 1/2\pi\sqrt{LC}$. The Q of the circuit is defined as $\omega_s L/R$ and affects the selectivity of the circuit through $\text{BW} = f_s/Q$. High-Q circuits are very selective. At resonance, $V_C \cong QE_{input}$. The half-power frequencies f_1 and f_2 are defined as those frequencies at which the power dissipated is one-half the power dissipated at resonance. In addition, the current is 0.707 (or $1/\sqrt{2}$) times the current at resonance.

The bandwidth $\text{BW} = f_2 - f_1$. The smaller the bandwidth, the more selective the circuit. In Fig. 12.1, note that increasing R results in a less selective circuit. Figure 12.2 shows the voltages across the three elements versus the frequency. The voltage across the resistor, V_R, has exactly the same shape as the current, since it differs by the constant R. V_R is a maximum at resonance. V_C and V_L are equal at resonance (f_s) since $X_L = X_C$, but note that they are not maximum at the resonant frequency. At frequencies below f_s, $V_C > V_L$; at frequencies above f_s, $V_L > V_C$, as indicated in Fig. 12.2.

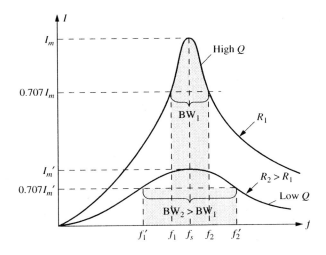

FIG. 12.1

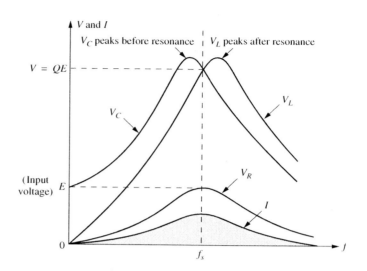

FIG. 12.2

PROCEDURE

Part 1 Low-Q Circuit

(a) Construct the circuit of Fig. 12.3. Insert the measured resistance values. In this experiment, the dc resistance of the coil (R_l) must be included. If available, the frequency counter should be employed.

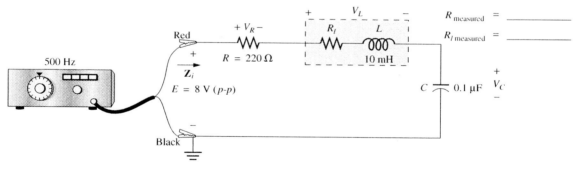

FIG. 12.3

(b) Using the nameplate values ($L = 10$ mH, $C = 0.1$ μF), compute the radian frequency ω_s and the frequency f_s at resonance. Show all your work in the following space. Organize the presentation and be neat! Record the results in Table 12.1.

TABLE 12.1

ω_s	
f_s	

Insert the calculated level of f_s in Table 12.2 and calculate the two additional frequency levels that are a function of f_s.

(c) Energize the circuit and set the oscillator to the frequencies indicated in Table 12.2. At each frequency, reset the input to 8 V (p-p) with the oscilloscope and measure the peak-to-peak values of the voltage V_C with the oscilloscope. **Take all the readings for V_C** and then interchange the positions of the capacitor and inductor and take all the V_L readings of Table 12.2. Be sure to take the additional readings near the peak value. Finally, interchange the positions of the inductor and resistor and take the V_R readings. This procedure must be followed to ensure a common ground between the source and measured voltage. Take a few extra readings near the resonant frequency. Use the dual-trace feature of the oscilloscope to measure the desired voltage and maintain E at 8 V (p-p) for each reading. If necessary, change scales to obtain the most accurate reading possible.

(d) Calculate the peak-to-peak value of the current at each frequency using Ohm's law and insert in Table 12.2.

(e) Calculate the input impedance at each frequency and complete Table 12.2.

(f) Plot Z_i versus frequency on Graph 12.1. Clearly indicate each plot point and label the curve and the resonant frequency.

TABLE 12.2

Frequency	$V_{C(p\text{-}p)}$	$V_{L(p\text{-}p)}$	$V_{R(p\text{-}p)}$	$I_{p\text{-}p} = \dfrac{V_{R(p\text{-}p)}}{R}$	$Z_i = \dfrac{E_{p\text{-}p}}{I_{p\text{-}p}}$
500 Hz					
1,000 Hz					
2,000 Hz					
3,000 Hz					
4,000 Hz					
5,000 Hz					
6,000 Hz					
7,000 Hz					
8,000 Hz					
9,000 Hz					
10,000 Hz					
$f_s = $ _____ Hz					
$f = 1.1f_s = $ _____					
$f = 0.9f_s = $ _____					
Additional readings near peak { _____ _____ _____					

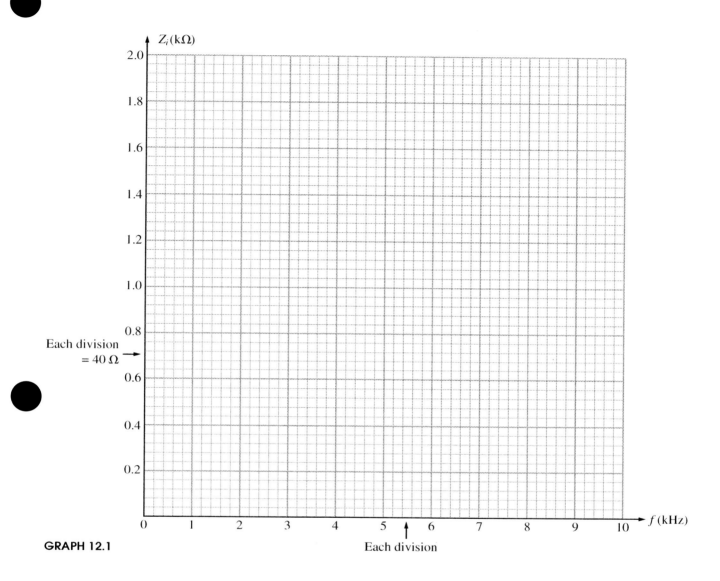

Each division = 40 Ω →

GRAPH 12.1

Each division

(g) At resonance, compare the input impedance Z_i to the total resistance of the circuit $R_T = R + R_l$. Record the values in Table 12.3.

TABLE 12.3

Z_i	
R_T	

(h) Describe in a few sentences how the input impedance of a series resonant circuit varies with frequency. Which element has the most influence on the input impedance at the low and high ends of the frequency spectrum?

(i) If the input impedance is a minimum at the resonant frequency, what would you expect to be true about the current at resonance?

Plot I_{p-p} versus frequency on Graph 12.2 and comment on whether the preceding conclusion was verified.

How does the maximum current compare to the value determined by $I_{p-p} = E_{p-p}/R_T = E_{p-p}/(R + R_l)$? Record both currents in Table 12.4.

Calculation:

TABLE 12.4

	Calculated	Graph
I_{max}		

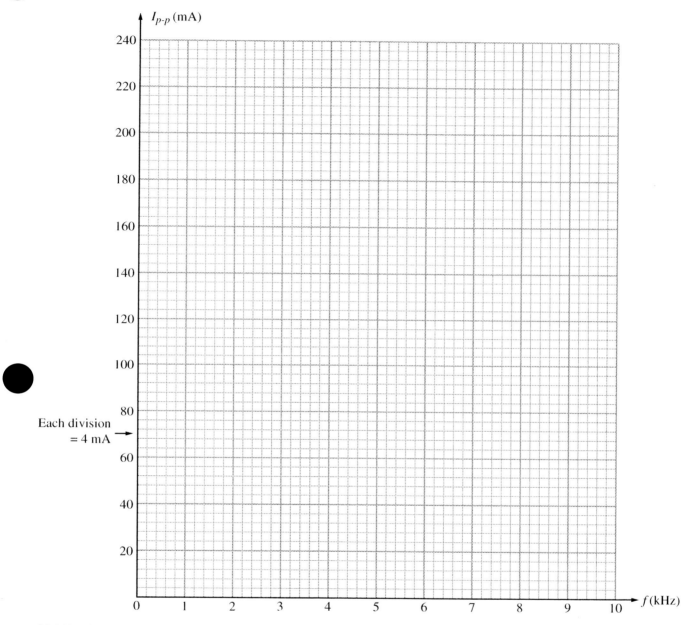

GRAPH 12.2

(j) Plot $V_{R(p\text{-}p)}$ versus frequency on Graph 12.3. Clearly indicate each point and label the curve. Use peak-to-peak values.

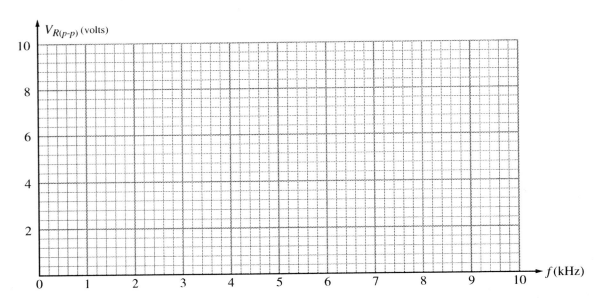

GRAPH 12.3

(k) Plot $V_{L(p\text{-}p)}$ and $V_{C(p\text{-}p)}$ versus frequency on Graph 12.4. Clearly indicate each plot point and label the curves. Use peak-to-peak values. Record the frequencies where V_R, V_L and V_C, are a maximum in Table 12.5.

TABLE 12.5

$f_{max}(V_R)$	
$f_{max}(V_L)$	
$f_{max}(V_C)$	
$f_1(\text{low})$	
$f_2(\text{high})$	
BW	
$Q_s(X/R)$	
$Q_s(f/BW)$	
f (in phase)	

Did V_C peak before f_s and V_L below f_s, as noted in the Résumé of Theory?

Does the maximum value of V_R occur at the same frequency noted for the current I? If so, why?

(l) On Graph 12.2, indicate the resonant and half-power frequencies and record in Table 12.5. Then determine the bandwidth and record in the same table.

Calculation:

Using measured resistor values, calculate the quality factor using the equation $Q_s = X_L/R_T$ and record in Table 12.5.

Calculation:

Then determine Q_s using the resonant frequency and bandwidth from Graph 12.2 and the equation $Q_s = f_s/BW$ and record in Table 12.5. Then compare with the result obtained.

Calculation:

Have the equations been verified? If not, why not?

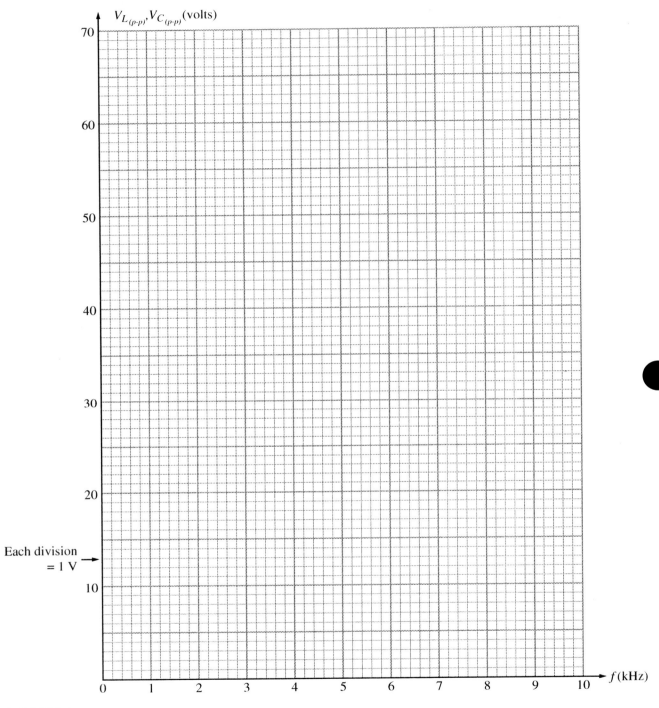

GRAPH 12.4

(**m**) After placing R in the position of C in Fig. 12.3, place E and V_R on a dual-trace oscilloscope. Vary the frequency applied from 0 to 10 kHz and note when E and V_R are in phase. At what frequency are the two in phase? Record the frequency in Table 12.5. When they are in phase, what does it reveal about the input impedance of the circuit at resonance?

Part 2   Higher-Q Circuit

We will now repeat the preceding analysis for a higher-Q (more selective) series resonant circuit by replacing the 220-Ω resistor with a 33-Ω resistor and noting the effect on the various plots.

(**a**) Measure the resistance of the 33-Ω resistor and record in Table 12.6. Repeat parts 1(a) through 1(e) after replacing the 220-Ω resistor with a 33-Ω resistor and enter the results in Tables 12.6 and Table 12.7. For readings near the maximum, you may have to use a 10:1 probe or set the GND line at the bottom of the display and multiply the resulting positive excursion by 2 to obtain the peak-to-peak value. Be sure to take a reading at the resonant frequency and the frequencies defined by f_s.

TABLE 12.6

R_{measured}	
Ω_s	
f_s	

(**b**) Plot Z_i versus frequency on Graph 12.1. Clearly indicate each plot point and label the curve to distinguish it from the other curve.

How has the shape of the curve changed?

Is the resonant frequency the same even though the resistance was changed?

Is the minimum value still equal to $R_T = R + R_l$?

TABLE 12.7

Frequency	$V_{R(p\text{-}p)}$	$V_{L(p\text{-}p)}$	$V_{C(p\text{-}p)}$	$I_{p\text{-}p} = \dfrac{V_{R(p\text{-}p)}}{R}$	$Z_i = \dfrac{E_{(p\text{-}p)}}{I_{(p\text{-}p)}}$
500 Hz					
1,000 Hz					
2,000 Hz					
3,000 Hz					
4,000 Hz					
5,000 Hz					
6,000 Hz					
7,000 Hz					
8,000 Hz					
9,000 Hz					
10,000 Hz					
$f_s = $ _____ Hz					
$f = 1.1, f_s = $ _____					
$f = 0.9, f_s = $ _____					
Additional readings near peak _____ _____ _____					

(c) Plot $I_{p\text{-}p}$ versus frequency on Graph 12.2. How has the shape of the curve changed with an increased Q value?

Is the maximum current the same, or has it changed?

Calculate the new maximum using the circuit values and compare to the measured graph value. Record the values in Table 12.8.

Calculation:

TABLE 12.8

	Calculated	Graph
I_{max}		

(d) Plot $V_{R(p-p)}$ versus frequency on Graph 12.3. Clearly indicate each plot point and label the curve. Use peak-to-peak values. Does the maximum value of V_R continue to occur at the same frequency noted for the current I? Record the frequency at which V_R is a maximum in Table 12.9.

TABLE 12.9

f_{max} (V_R)	
f_{max} (V_L)	
f_{max} (V_C)	
f_1 (low)	
f_2 (high)	
BW	
Q_s (X/R)	
Q_s (f/BW)	

(e) Plot $V_{L(p-p)}$ and $V_{C(p-p)}$ versus frequency on Graph 12.4. Clearly indicate each plot point and label the curves. Use peak-to-peak values. Record the frequencies where V_L and V_C are maximum values in Table 12.9.

Are the frequencies at which V_L and V_C reached their maximums closer to the resonant frequency than they were for the low-Q network? A theoretical analysis will reveal that the higher the Q of the network, the closer the maximums of V_L and V_C are to the resonant frequency.

(f) On Graph 12.2, indicate the half-power frequencies and the bandwidth and record the values in Table 12.9.

Calculation:

Using measured resistor values, calculate the quality factor using the equation $Q_s = X_L/R_T$ and record in Table 12.9.

Calculation:

Then calculate Q_s using the resonant frequency and bandwidth from Graph 12.2 and the equation $Q_s = f_s/\text{BW}$ and record in Table 12.9 compare with the result obtained above.

Calculation:

Have the equations been verified? If not, why not?

EXERCISES

1. Design a series resonant circuit with the following specifications:

$L = 10$ mH, $\qquad R_l = 4\ \Omega$ $\qquad I_{max} = 200$ mA

$f_s = 10$ kHz, $\qquad Q_s = 20$

That is, determine the required R, C, and supply voltage E. For R and C, use the closest standard values. Consult any catalog in the laboratory area for standard values.

$R =$ _____, $C =$ _____, $E =$ _____

2. Redesign the network of Problem 1 to have a Q_s of 10. All the other specifications remain the same.

$R =$ _____, $C =$ _____, $E =$ _____

3. Test the design of problem 1 using PSpice or Multisim. That is, once you have the design, obtain a plot of I versus frequency and note whether I_{max} is close to 200 mA and the bandwidth close to 500 Hz. Attach all pertinent printouts.

Name:_____

Date:_____

Course and Section:_____

Instructor:_____

Parallel Resonant Circuits

OBJECTIVES

1. Validate the basic equations for the resonant frequency of a parallel resonant circuit.
2. Plot the voltages and source current (constructed) versus frequency.
3. Demonstrate how the input impedance varies with frequency.
4. Validate the relationship between the quality factor of a network and the bandwidth.

EQUIPMENT REQUIRED

Resistors

1—4.7-Ω, 47-Ω, 100-kΩ (1/4-W)

Inductors

1—10-mH

Capacitors

1—0.1-μF

Instruments

1—DMM

1—Oscilloscope

1—Audio oscillator or function generator

1—Frequency counter (strongly recommended)

EQUIPMENT ISSUED

TABLE 13.0

Item	Manufacturer and Model No.	Laboratory Serial No.
DMM		
Oscilloscope		
Audio oscillator or function generator		
Frequency counter		

RÉSUMÉ OF THEORY

The basic components of a parallel resonant network appear in Fig. 13.1.

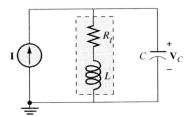

FIG. 13.1

For parallel resonance, the following condition must be satisfied:

$$\frac{1}{X_C} = \frac{X_L}{X_L^2 + R_l^2} \tag{13.1}$$

At resonance, the impedance of the network is resistive and is determined (for $Q_l \geq 10$) by

$$Z_{Tp} = \frac{L}{R_l C} \tag{13.2}$$

The resonant frequency is determined by

$$f_p = \frac{1}{2\pi\sqrt{LC}}\sqrt{1 - \frac{R_l^2 C}{L}} \tag{13.3}$$

which reduces to

$$f_p = \frac{1}{2\pi\sqrt{LC}} \tag{13.4}$$

when

$$Q_{\text{coil}} = \frac{X_L}{R_l} \geq 10$$

For parallel resonance, the curve of interest is that of $\mathbf{V}_C$ versus frequency due to electronic considerations that often place this voltage at the input to the following stage. It has the same shape as the resonance curve for the series configuration with the bandwidth determined by

$$\boxed{\text{BW} = f_2 - f_1} \tag{13.5}$$

where f_2 and f_1 are the cutoff or band frequencies.

PROCEDURE

Part 1 High-Q Parallel Resonant Circuit

(a) Construct the network of Fig. 13.2. Insert the measured resistance values.

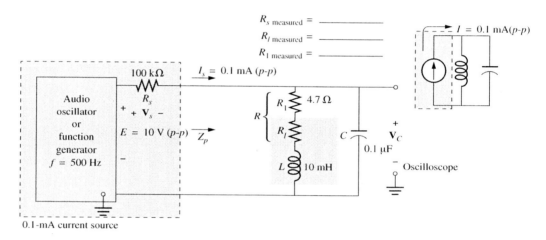

R_s measured = _____

R_l measured = _____

R_1 measured = _____

$I = 0.1$ mA(p-p)

$I_s = 0.1$ mA (p-p)

0.1-mA current source

FIG. 13.2

To limit the quality factor of the circuit to one that can be easily graphed using the linear scales appearing in this laboratory exercise, a resistor R_1 was added in series with R_l to increase the total resistance in series with the coil to $R = R_1 + R_l$. Using measured values, determine R and record in Table 13.1. For all the calculations to follow, be sure to insert R in place of R_l as appearing in Eqs. (13.1)–(13.3). This is the manner in which the shape of a resonant curve can be controlled. By inserting a potentiometer in series with the inductor, an infinite variety of shapes is available.

(b) The input voltage E must be maintained at 10 V (p-p). Make sure to reset E after each new frequency is set on the oscillator. This is a perfect opportunity to use the dual-trace feature of the oscilloscope to measure the desired voltage and maintain E at 10 V (p-p).

Since the input to a parallel resonant circuit is usually a constant-current device, such as a transistor, the input current I_s has been designed to be fairly constant in magnitude for the frequency range of interest. The input impedance (Z_p) of the parallel resonant circuit will always be sufficiently less than the 100-$k\Omega$ resistor, to permit the following approximation:

$$I_s = \frac{E}{100\ k\Omega + Z_p} \cong \frac{E}{100\ k\Omega} = \frac{10\ \text{V }(p\text{-}p)}{100\ k\Omega} = 0.1\ \text{mA }(p\text{-}p)$$

(c) Compute the resonant frequency of the network of Fig. 13.2 using the nameplate data, measured resistor values, and Eq. (13.3). Be certain to insert R in place of R_l. Record in Table 13.1.

Calculation:

TABLE 13.1

R	
f_p (Eq. 13.3)	
f_p (Eq. 13.4)	
Z_p	
Q_l	
Q_p	

(d) Compute the resonant frequency of the network of Fig. 13.2 using the nameplate data, measured resistor values, and Eq. (13.4). Record the result in Table 13.1. How do the two levels of f_p compare? Can we assume $Q_l \geq 10$ for this configuration? Use the value determined in part 1(c) for all future calculations.

Calculation:

(e) Calculate the input impedance (the maximum value) at resonance using the following equation:

$$Z_p = R_p = \frac{L}{RC}$$

and record in Table 13.1.

Calculation:

(**f**) Compare the 100-kΩ resistor with the result of part 1(c). Is it reasonable to assume that the input current is fairly constant for the frequency range of interest?

(**g**) Is this a high- or low-Q circuit? First calculate Q_l at resonance using

$$Q_l = \frac{X_L}{R}$$

and record in Table 13.1.

Calculation:

If $Q_l \geq 10$, $Q_p \cong Q_l$. If this is the case, record Q_p in Table 13.1.

(**h**) Vary the input frequency from 500 Hz to 10 kHz and use the oscilloscope to measure $V_{C_{(p\text{-}p)}}$ at each frequency. Use the dual-trace feature of the oscilloscope to measure $V_{C_{(p\text{-}p)}}$ and maintain E at 10 $V_{(p\text{-}p)}$. (You may find as the frequency increases that the peak-to-peak value of E will begin to drop from the initial 10 $V_{(p\text{-}p)}$ setting). Change scales to obtain the most accurate reading at each frequency. Insert the results in Table 13.2 taking a few extra readings at and near the resonant frequency. Use the fine control on your generator to obtain an accurate reading for the resonant frequency.

Then move R_s as indicated in Fig. 13.3 (to maintain a common ground between the generator and oscilloscope), measure $V_{R_{s(p\text{-}p)}}$ at each frequency of Table 13.2, and insert in

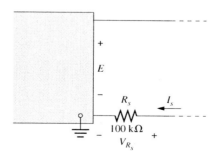

FIG. 13.3

Table 13.2, all the while maintaining E at 10 $V(p\text{-}p)$. Calculate $I_{s(p\text{-}p)}$ from $I_{s(p\text{-}p)} = V_{Rs(p\text{-}p)}/R_s$ at each frequency and enter in Table 13.2. Again, take a few extra readings at and near the resonant frequency. Be accurate and neat!

The impedance of a parallel resonant network (Z_p) can be determined by the ratio of the applied voltage (across the resonant circuit) divided by the source current. Since all branches of a parallel resonant network are in parallel, the applied voltage is the same as the voltage across the capacitor. The result is that the impedance Z_p can be determined from $Z_p = V_{C(p\text{-}p)}/I_{s(p\text{-}p)}$. Calculate Z_p at each frequency of Table 13.2 and complete the table.

TABLE 13.2

Frequency	$V_{C(p\text{-}p)}$	$V_{Rs(p\text{-}p)}$	$I_{s(p\text{-}p)}$	$Z_p = \dfrac{V_{C(p\text{-}p)}}{I_{c(p\text{-}p)}}$
500 Hz				
1,000 Hz				
2,000 Hz				
3,000 Hz				
4,000 Hz				
5,000 Hz				
6,000 Hz				
7,000 Hz				
8,000 Hz				
9,000 Hz				
10,000 Hz				
(at peak) $f_p =$ _____ Hz				
(below peak) $f =$ _____ Hz				
(below peak) $f =$ _____ Hz				
(above peak) $f =$ _____ Hz				
(above peak) $f =$ _____ Hz				

(i) Plot Z_p versus frequency on Graph 13.1 using the data of Table 13.2. Be sure to indicate each plot point clearly and label the resulting plot. Choose a vertical scale that will best display the resulting data.

Is the network impedance a minimum or maximum at resonance?

What was the relative value of the input impedance of a series resonant circuit at resonance?

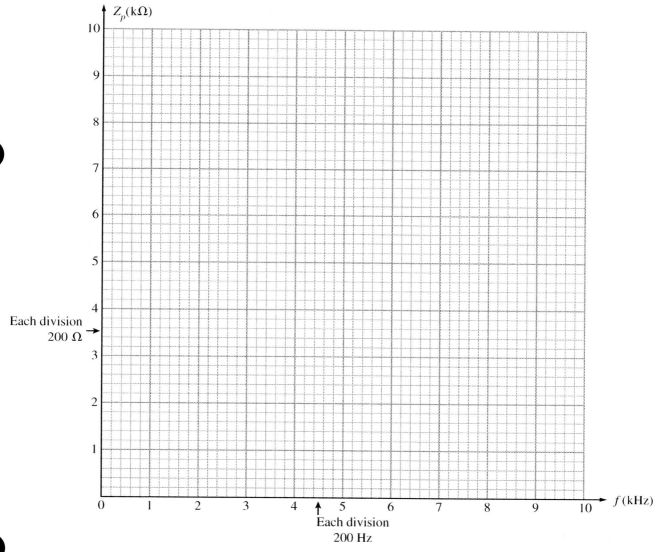

GRAPH 13.1

What is the maximum value of Z_p? How does it compare to the calculated value of part 1(e)? Record the calculated and measured value in Table 13.3.

TABLE 13.3

	Calculated	Measured
Z_p		
F_p		

(j) Plot $I_{s(p-p)}$ versus frequency on Graph 13.2 using the data of Table 13.2. Is $I_{s(p-p)}$ fairly constant for the frequency range of interest? Are the characteristics of the current source satisfactory for the type of analysis being performed?

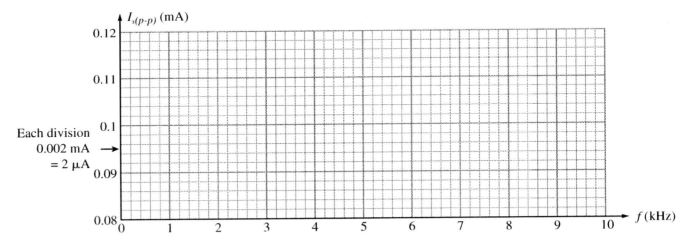

GRAPH 13.2

(k) Plot $V_{C(p-p)}$ versus frequency on Graph 13.3 using the data of Table 13.2. Again, choose the vertical scale that will best display the resulting data. Clearly identify each plot point and label the curve. Find the resonant frequency and define the bandwidth at the $0.707V_{C(p-p)}$ level. Drop down to the horizontal axis to obtain and label the low and high cutoff frequencies.

What is the resonant frequency f_p? How does it compare to the calculated level of part 1(c)? Record the calculated and measured value in Table 13.3.

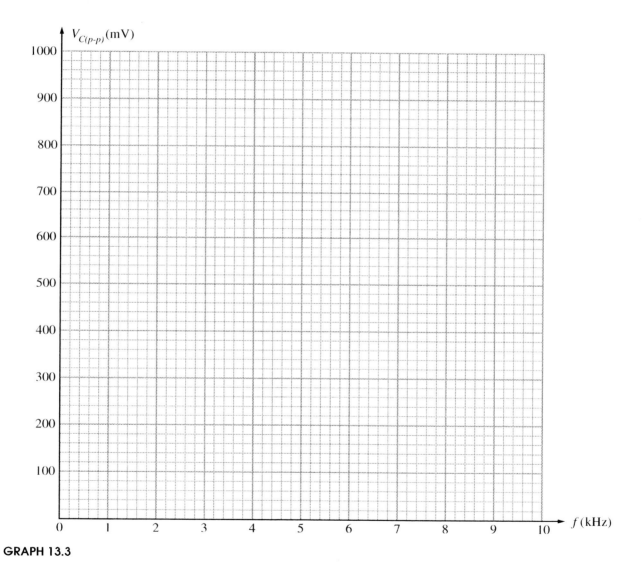

GRAPH 13.3

What are f_1 (low cutoff), f_2 (high cutoff), and the bandwidth for this design? Record the results in Table 13.4.

Compute the bandwidth from BW = f_p/Q using calculated values from parts 1(c) and 1(g) and compare with the preceding result. Record in Table 13.4.

Calculation:

TABLE 13.4

f_1 (low)	
f_2 (high)	
BW (graph)	
BW (f/Q)	
Q_p	

Using the levels determined from the graph, calculate Q_p from $Q_p = f_p/\text{BW}$ and compare to the calculated level of part 1(g). Record in Table 13.4.

Calculation:

Part 2   Lower-Q Parallel Resonant Circuit

Let us now examine the impact of a lower quality factor Q on the shape of the resonant curves. The quality factor will be reduced for the network of Fig. 13.2 by replacing the 4.7 Ω resistor with a 47 Ω resistor in series with the inductor to increase the resistance of the branch to $R = 47\ \Omega + R_l$.

(a) Reconstruct the network of Fig. 13.2 with a 47-Ω resistor in series with the inductor.

Determine $R = 47\ \Omega + R_l$ using measured values and record Table 13.5.
For all calculations to follow, be sure to use R in place of R_l in all applicable equations.

TABLE 13.5

R	
f_p	
Z_p	
Q_l	
Q_p	

(b) Determine the resonant frequency using Eq. 13.3 and record in Table 13.5.

Calculation:

Has the increase in resistance affected the resonant frequency? That is, how does this frequency compare to that obtained in part 1(c)?

(c) Calculate the input impedance at resonance using the equation $Z_p = \dfrac{L}{RC}$ and record in Table 13.5.

How does it compare to the level calculated in part 1(e)?

Do you expect the characteristics of the constructed current source to improve or deteriorate? Why?

(d) Is this a high- or low-Q circuit? First find Q_l using $Q_l = X_L/R_T$ and record in Table 13.5.

Calculation:

If $Q_l \leq 10$, use the following equation to determine Q_p:

$$Q_p = \frac{Z_{T_p}}{X_C} = \frac{\dfrac{(R^2 + X_L^2)}{R}}{X_C} = \frac{R^2 + X_L^2}{RX_C} \qquad (13.6)$$

where Z_{T_p} is the magnitude of the total impedance of the parallel resonant network at resonance. Record the result in Table 13.5.

How does Q_p compare to Q_l?

(e) Repeat part 1(h) and insert the data in Table 13.6.

TABLE 13.6

Frequency	$V_{C(p\text{-}p)}$	$V_{R_{s(p\text{-}p)}}$	$I_{s(p\text{-}p)}$	$Z_p = \dfrac{V_{C(p\text{-}p)}}{I_{s(p\text{-}p)}}$
500 Hz				
1,000 Hz				
2,000 Hz				
3,000 Hz				
4,000 Hz				
5,000 Hz				
6,000 Hz				
7,000 Hz				
8,000 Hz				
9,000 Hz				
10,000 Hz				
(at peak) $f_p =$ _____ Hz				
(below peak) $f =$ _____ Hz				
(below peak) $f =$ _____ Hz				
(above peak) $f =$ _____ Hz				
(above peak) $f =$ _____ Hz				

(f) Calculate Z_p at each frequency and complete the table.

Plot Z_p versus frequency on Graph 13.1 using the data of Table 13.2. Clearly indicate each plot point and label the curve.

How does this curve of Z_p versus frequency compare to that with the higher Q?

What is the maximum value of Z_p? How does it compare to the calculated value of part 2(c)? Record the calculated and measured values in Table 13.7.

TABLE 13.7

	Calculated	Measured
Z_p		
f_p		

(g) Plot $I_{s(p\text{-}p)}$ versus frequency on Graph 13.2 using the data of Table 13.2 and label the curve to distinguish it from the first plot. Is $I_{s(p\text{-}p)}$ fairly constant for the frequency range of interest? Is it an improved set of characteristics as compared to that obtained for the high-Q network? If so, why?

(h) Plot $V_{C(p\text{-}p)}$ versus frequency on Graph 13.3 using the data of Table 13.3. Clearly identify each plot point and label the curve. Find the resonant frequency and identify the bandwidth at the $0.707V_{C(p\text{-}p)}$ level. Drop down to the horizontal axis to obtain and label the low and high cutoff frequencies.

What is the resonant frequency f_p? How does it compare to the calculated level of part 2(b)? Record both levels in Table 13.7.

What are f_1 (low cutoff), f_2 (high cutoff), and the bandwidth for this design? Record their values in Table 13.8.

Compute the bandwidth from BW $= f_p/Q_p$ using calculated values from parts 2(b) and 2(d) and record in Table 13.8. Compare with the preceding result. Explain any major differences.

Calculation:

TABLE 13.8

f_1 (low)	
f_2 (high)	
BW (graph)	
BW (f/Q)	
Q_p	

Using the levels determined from the graph, calculate Q_p from $Q_p = f_p/\text{BW}$ and record the result in Table 13.8. Compare to the calculated level of part 2(d). Explain any major differences.

Calculation:

EXERCISES

1. If the capacitor of Fig. 13.2 were decreased to 0.01 μF, would the current source be a good design for the chosen frequency range?

2. Determine the approximate bandwidth of a parallel resonant circuit with $I_s = 5$ mA, $R_s = \infty\,\Omega$, $L = 100\ \mu H$, $R_l = 5\ \Omega$, and $C = 0.02\ \mu F$.

BW = _____

3. Using PSpice or Multisim, obtain a plot of V_C vs. frequency for the network of Fig. 13.2 and compare with the results plotted on Graph 13.3. Attach all pertinent printouts.

EXPERIMENT ac **14**

Passive Filters

OBJECTIVES

1. Become familiar with the characteristics of passive low-pass and high-pass filters.
2. Plot the phase response for low-pass and high-pass filters.
3. Analyze the frequency response of tuned band-pass and band-stop filters.
4. Verify a number of equations and conclusions associated with filter networks.
5. Become adept in the use of semilog graph paper.

EQUIPMENT REQUIRED

Resistors

1—100-Ω, 1-kΩ, 22-kΩ

Capacitors

1—0.001-μF, 0.1-μF, 0.2-μF

Inductors

1—1-mH, 10-mH

Instruments

1—DMM

1—Oscilloscope

1—Audio oscillator or function generator

1—Frequency counter (strongly recommended)

EQUIPMENT ISSUED

TABLE 14.0

Item	Manufacturer and Model No.	Laboratory Serial No.
DMM		
Oscilloscope		
Audio oscillator or function generator		
Frequency counter		

RÉSUMÉ OF THEORY

A filter is a device that will pass signals of certain frequencies while rejecting signals of other frequencies. There are two types:

1. Passive filters are those made up of resistors, capacitors, and inductors but no active devices such as transistors, etc.
2. Active filters are made up of resistors and capacitors, but they use active devices in circuits called operational amplifiers that simulate the inductor in the filter.

In this laboratory exercise, we will study the characteristics of low-pass, high-pass, band-pass, and band-stop passive filters.

If the band of frequencies passed by a filter begins at zero and increases up to a particular frequency above which all others are rejected, the filter is called a *low-pass filter,* because it passes the low frequencies and rejects those above a certain value.

On the other hand, if the filter rejects all frequencies from zero up to a particular frequency and passes those above this frequency, the filter is called a *high-pass filter.* Figure 14.1

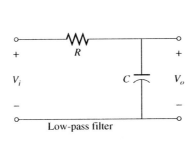

Low-pass filter

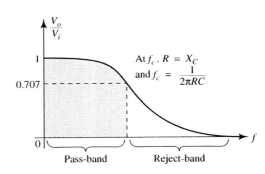

$$\text{At } f_c, R = X_C$$
$$\text{and } f_c = \frac{1}{2\pi RC}$$

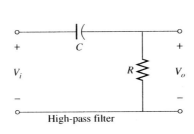

High-pass filter

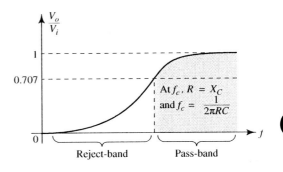

$$\text{At } f_c, R = X_C$$
$$\text{and } f_c = \frac{1}{2\pi RC}$$

FIG. 14.1

shows the circuits, defining equations, and the frequency response of both high- and low-pass filters.

The frequency that delineates between passage and rejection is called the *cutoff frequency*. In the case of the high-pass filter, it is the low-frequency cutoff, whereas in the case of the low-pass filter, it is called the high-frequency cutoff. This frequency is defined as the frequency at which the output voltage drops to 0.707 of its maximum value and at which $R = X_C$ in the circuit.

If both of these circuits exist simultaneously, as they do in an R-C coupled amplifier, then the result is a pass-band response, as shown in Fig. 14.2. The band of frequencies between the high and low cutoff frequencies is called the bandwidth (BW) and is defined as shown.

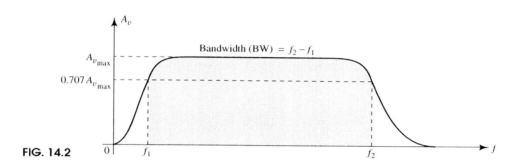

FIG. 14.2

Tuned filters use all three components—resistors, capacitors, and inductors—to achieve their goals. These filter circuits rely heavily on the operational concept known as resonance. A circuit is said to be at resonance when the total stored energy in the magnetic field of the inductor is transferred to the capacitor, which stores it in the form of an electric field. The capacitor then discharges so that the energy is again stored in the magnetic field of the inductor. At the *resonant frequency*, the reactances are equal ($X_L = X_C$) and the circuit is purely resistive.

A number of tuned filter circuits appear in Fig. 14.3, which continues on the next page.

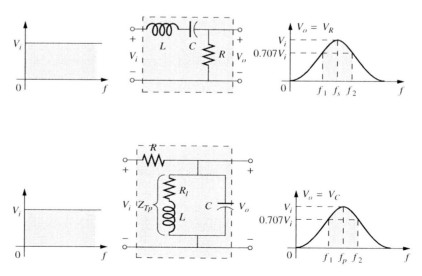

FIG. 14.3   (a)

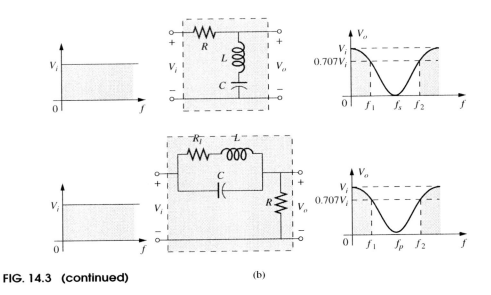

FIG. 14.3   (continued)   (b)

PROCEDURE

Part 1   High-pass *R-C* Filter

(a) Construct the network of Fig. 14.4. Insert the measured value of the resistor *R*.
Note the common ground of the generator and channels 1 and 2 of the oscilloscope.

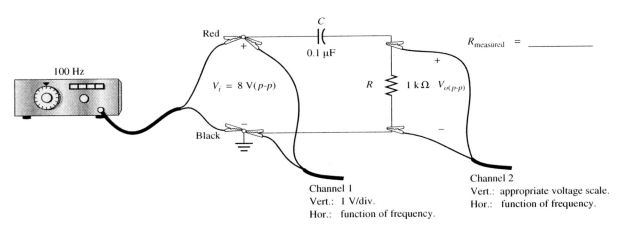

FIG. 14.4

(b) Write an equation for $\mathbf{V}_o$ in terms of $\mathbf{V}_i$.

$$\mathbf{V}_o = \underline{\hspace{3cm}}$$

Then establish the ratio $\mathbf{V}_o/\mathbf{V}_i$ in magnitude and phase form.

$$\mathbf{V}_o/\mathbf{V}_i = \underline{\hspace{2cm}}$$

The condition $R = X_C$ establishes the low cutoff frequency. Write an equation for the frequency that establishes this condition. Show the derivation.

$$f_c = \underline{\hspace{2cm}}$$

Using the preceding equation and the measured resistor value, solve for the low cut-off frequency and record in Table 14.1.

Calculation:

TABLE 14.1

	Calculated	Graph	Graph (45°)
f_c			
V_o/V_i (2 kHz)			

(c) Energize the network of Fig. 14.4 and set the signal generator to each of the frequencies appearing in Table 14.2. For each setting of the generator, make sure that $V_i = 8$ V (p-p). Measure V_o (p-p) and calculate the ratio A_v to complete Table 14.2.

TABLE 14.2

Frequency (kHz)	$V_{o_{(p\text{-}p)}}$	$A_v = \dfrac{V_{o_{(p\text{-}p)}}}{V_{i_{(p\text{-}p)}}}$
0.1		
0.2		
0.4		
0.6		
0.8		
1.0		
1.2		
1.4		
1.6		
1.8		
2.0		
3.0		
4.0		
5.0		
6.0		
8.0		
10.0		
12.0		
14.0		
16.0		
18.0		
20.0		
40.0		
60.0		
100.0		

(d) Plot A_v versus frequency on Graph 14.1. Clearly identify each plot point and label the curve.

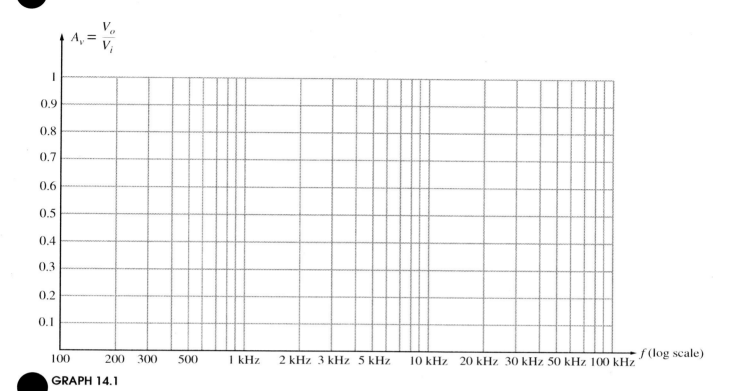

GRAPH 14.1

(e) Draw a horizontal dashed line at the 0.707 level and determine the frequency at which it intersects the curve. Record the low cutoff frequency in Table 14.1.

How does it compare to the predicted level calculated in part 1(b)?

(f) Using the equation derived in part 1(b), calculate the magnitude of the ratio V_o/V_i at $f = 2$ kHz and record in Table 14.1.

Calculation:

Using the curve of Graph 14.1, determine the ratio at $f = 2$ kHz and record in Table 14.1. Compare to the calculated level.

(g) Calculate the phase angle of V_o/V_i using the equation derived in part 1(b) at the frequencies appearing in Table 14.3. The angle θ is the angle by which V_o leads the input voltage V_i.

TABLE 14.3

Frequency (kHz)	θ
0.1	
0.2	
0.6	
1	
2	
6	
10	
12	
20	
40	
60	
100	

(h) Plot θ versus frequency on Graph 14.2. Clearly indicate each plot point and label the curve.

(i) Determine the frequency at which the angle θ is 45° and record in the top row of Table 14.1.

How close is it to the low cutoff frequency calculated in part 1(b)?

(j) Based on the above results, is the high-pass filter a leading or lagging network? Why?

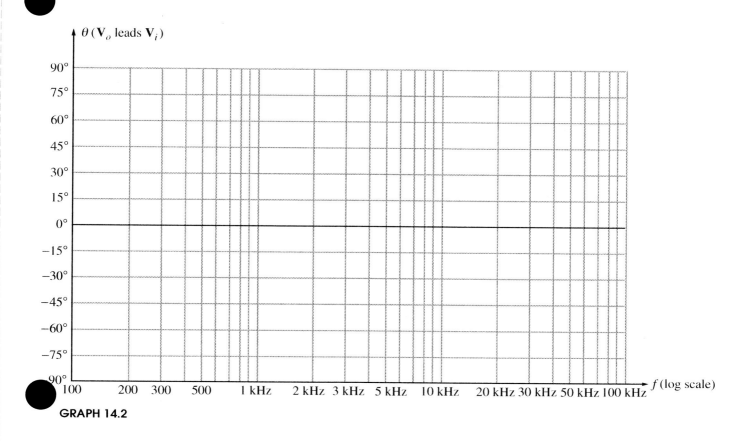

GRAPH 14.2

Part 2 Tuned Band-Pass Filter

(a) Construct the network of Fig. 14.5. Insert the measured value of the resistor R. Measure R_l for the inductor and record in the figure.

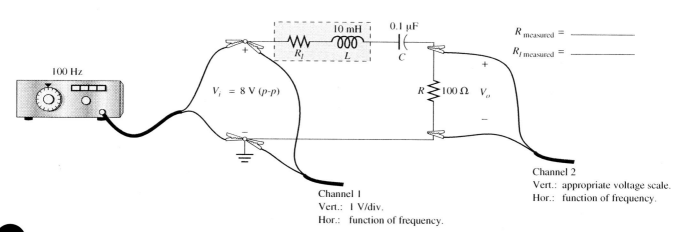

FIG. 14.5

(**b**) Calculate the resonant frequency of the series resonant circuit and record in Table 14.4. Show all work.

Calculation:

TABLE 14.4

f_s	
Q_s	
BW (f/Q)	
V_o/V_i (at f_s)	
BW	
$A_{v_{max}}$	
f_i (low)	
f_i (high)	

(**c**) Using the calculated resonant frequency, calculate the quality factor of the network and record in Table 14.4.

Calculation:

(**d**) Using the results of parts 3(b) and 3(c), calculate the bandwidth of the resonant circuit and record in Table 14.4.

Calculation:

(e) If the bandwidth as calculated in part 3(d) is for the current through the series circuit, will the voltage V_o have the same bandwidth? Why?

(f) Determine the magnitude of the voltage V_o at resonance using measured resistor values. Then determine the ratio V_o/V_i at resonance and record in Table 14.4.

Calculation:

(g) Energize the network of Fig. 14.5 and set the signal generator to each of the frequencies appearing in Table 14.5. For each setting of the generator, be sure that $V_i = 8$ V (*p-p*). Measure V_o (*p-p*) and calculate the ratio A_v to complete Table 14.5.

(h) Plot A_v versus frequency on Graph 14.3. Clearly indicate each plot point and label the curve.

(i) Draw a horizontal dashed line at 0.707 ($A_{v(max)}$) to define the bandwidth of the filter. Record the bandwidth in Table 14.4. How does it compare to the calculated level of part 3(d)?

(j) Find the maximum value of A_v and record in Table 14.4. Compare to the calculated level of part 3(f).

TABLE 14.5

Frequency (kHz)	$V_{o_{(p\text{-}p)}}$	$A_v = \dfrac{V_{o_{(p\text{-}p)}}}{V_{i_{(p\text{-}p)}}}$
0.1		
0.2		
0.4		
0.6		
0.8		
1.0		
1.2		
1.4		
1.6		
1.8		
2.0		
3.0		
4.0		
5.0		
6.0		
8.0		
10.0		
12.0		
14.0		
16.0		
18.0		
20.0		
40.0		
60.0		
100.0		

A few
more
near
resonance

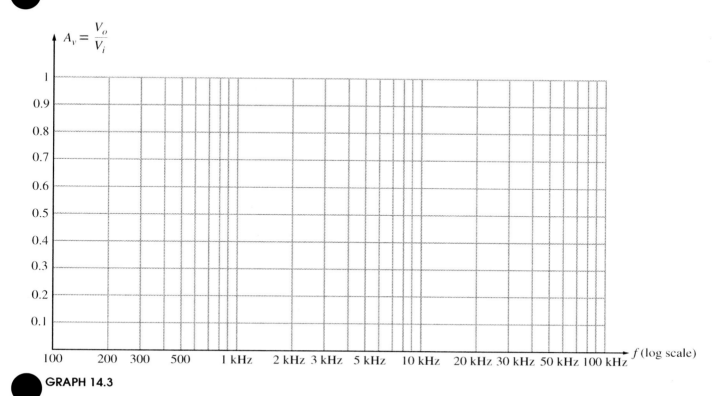

$$A_v = \frac{V_o}{V_i}$$

f (log scale)

GRAPH 14.3

(k) Record the upper and lower cutoff frequencies as defined at the 0.707 ($A_{v(max)}$) level in Table 14.4. What range of frequencies will pass through the filter with a measurable amount of power?

Part 3   Tuned Band-Stop Filter

(a) Construct the network of Fig. 14.6. Again insert the measured resistor values.

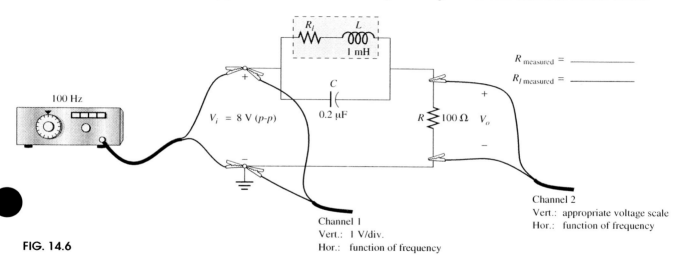

$R_{measured} = $ _____

$R_{l\ measured} = $ _____

Channel 1
Vert.: 1 V/div.
Hor.:   function of frequency

Channel 2
Vert.: appropriate voltage scale
Hor.:   function of frequency

FIG. 14.6

(b) Calculate the resonant frequency of the parallel resonant network and record in Table 14.6.

Calculation:

TABLE 14.6

f_p	
Q_l	
BW (Calc.)	
Z_{T_P}	
V_o/V_i(Calc.)	
BW (graph)	
V_o/V_i (graph)	
f_1 (low)	
f_2 (high)	

(c) Using the calculated resonant frequency, determine the quality factor of the coil and record in Table 14.6.

Calculation:

(d) Using the results of parts 4(b) and 4(c), calculate the bandwidth of the resonant circuit and record in Table 14.6.

Calculation:

(e) Would you expect the bandwidth of the tuned network to be exactly the same as that for V_o? If so, why? If not, why not? A brief sentence or two is sufficient.

(f) Determine the impedance at resonance of the parallel resonant network and record in Table 14.6. Calculate the magnitude of the voltage V_o. Then determine the ratio V_o/V_i and record in Table 14.6.

Calculation:

(g) Energize the network of Fig. 14.7 and set the signal generator to each of the frequencies appearing in Table 14.7. For each setting of the generator, be sure that $V_i = 8$ V (p-p). Measure V_o (p-p) and calculate the ratio A_v to complete Table 14.7.

(h) Plot A_v versus frequency on Graph 14.3. Clearly identify each plot point and label the curve.

(i) Draw a horizontal dashed line at 0.707 ($A_{v(max)}$) to define the bandwidth of the filter and record in Table 14.6. How does it compare to the calculated level of part 3(d)? Does the result support or oppose your earlier conclusions?

(j) Find the ratio V_o/V_i at resonance and compare to the level calculated in part 4(f).

(k) Record the upper- and lower-band frequencies as defined by the 0.707 ($A_{v(max)}$) level in Table 14.6. What range of frequencies will be blocked by the filter?

TABLE 14.7

Frequency (kHz)	$V_{o_{(p-p)}}$	$A_v = \dfrac{V_{o_{(p-p)}}}{V_{i_{(p-p)}}}$
0.1		
0.2		
0.4		
0.6		
0.8		
1.0		
1.2		
1.4		
1.6		
1.8		
2.0		
3.0		
4.0		
5.0		
6.0		
8.0		
10.0		
12.0		
14.0		
16.0		
18.0		
20.0		
40.0		
60.0		
100.0		
A few more near resonance		

EXERCISES

1. Using $E = 8$ V (p-p) $\angle 0°$, determine $\mathbf{V}_o$ at $f = f_p$ for the network of Fig. 14.6. That is, determine X_L and X_C at the designated frequency and use series-parallel techniques to determine $\mathbf{V}_o$ (magnitude and phase). Use peak-to-peak values for all the calculations. Then determine the magnitude of the ratio $\mathbf{V}_o/\mathbf{V}_i$ and compare to the value determined in part 4(j). Record the phase angle of the ratio $\mathbf{V}_o/\mathbf{V}_i$ and comment on whether it is close to what you expected at resonance.

$V_o/V_i =$ _____ , $\theta =$ _____

2. Develop an equation for the phase relationship between $\mathbf{V}_o$ and $\mathbf{V}_i$ for the network of Fig. 14.5—that is, an equation for the angle by which $\mathbf{V}_o$ leads $\mathbf{V}_i$. Then determine the phase angle at $f = f_s$, $\frac{1}{2}f_s$, and $2f_s$ and comment on whether they are the phase angles you expected at resonance and at the lower and higher frequencies.

3. Using PSpice or Multisim, generate a plot of V_o vs frequency for the network of Fig. 14.6. Use a frequency range of 100 Hz to 100 kHz and compare the resulting plot to that appearing in Graph 14.3. Attach all pertinent printouts.

Name:_____

Date:_____

Course and Section:_____

Instructor:_____

EXPERIMENT ac **15**

The Transformer

OBJECTIVES

1. Examine the phase relationship between the transformer primary and secondary voltages.
2. Measure the voltages and currents of a transformer and verify the impact of the turns ratio on their magnitudes.
3. Note the relationship between the apparent power levels of the primary and secondary circuits.
4. Verify the basic equation for the input impedance of a transformer under loaded conditions.

EQUIPMENT REQUIRED

Resistors

1—10-Ω, 100-Ω, 220-Ω, 1-kΩ (1/4-W)

Capacitors

1—100-μF

Transformers

Filament—120-V primary, 12.6-V secondary

Instruments

1—DMM

1—Oscilloscope

1—Audio oscillator or function generator

EQUIPMENT ISSUED

TABLE 15.0

Item	Manufacturer and Model No.	Laboratory Serial No.
DMM		
Oscilloscope		
Audio oscillator or function generator		

RÉSUMÉ OF THEORY

A *transformer* is a device that transfers energy from one circuit to another by electromagnetic induction. The energy is always transferred without a change in frequency. The winding connected to the energy source is called the *primary,* while the winding connected to the load is called the *secondary.* A step-up transformer receives electrical energy at one voltage and delivers it at a higher voltage. Conversely, a step-down transformer receives energy at one voltage and delivers it at a lower voltage. Transformers require little care and maintenance because of their simple, rugged, and durable construction. The efficiency of power transformers is also quite high.

The operation of the transformer is based on the principle that electrical energy can be transferred efficiently by mutual induction from one winding to another.

The physical construction of a transformer is shown in Fig. 15.1.

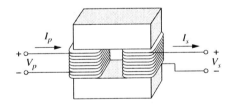

FIG. 15.1

For practical applications, the apparent power at the input to the primary circuit $(S = V_p I_p)$ is equal to the apparent power rating at the secondary $(S = V_s I_s)$. That is,

$$\boxed{V_p I_p = V_s I_s}$$

(15.1)

It can also be shown that the ratio of the primary voltage V_p to the secondary voltage V_s is equal to the ratio of the number of turns of the primary to the number of turns of the secondary:

$$\boxed{\frac{V_p}{V_s} = \frac{N_p}{N_s}}$$

(15.2)

The currents in the primary and secondary circuits are related in the following manner:

$$\boxed{\frac{I_p}{I_s} = \frac{N_s}{N_p}}$$

(15.3)

The impedance of the secondary circuit is electrically reflected to the primary circuit as indicated by the following expression:

$$\boxed{Z_p = a^2 Z_s}$$ (15.4)

where $a = N_p/N_s$

For safety reasons, the applied voltage and resulting current levels will be significantly less than normally applied to a filament transformer. The result is that Eqs. (15.1)–(15.4) may not be substantiated on an exact basis.

PROCEDURE

Part 1   Phase Relationship between the Transformer Primary and Secondary Voltages

(a) Set up the circuit of Fig. 15.2. Note that the primary voltage is 12 V (p-p).

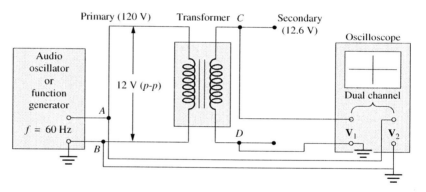

FIG. 15.2

Using the alternate mode, adjust the vertical sensitivities so that both deflections are equal. What does the pattern on the screen indicate as to the phase relationship of the two voltages?

(b) Calculate the turns (transformation) ratio from $a = N_p/N_s = V_p/V_s = 12$ V (p-p)/ $V_s(p\text{-}p) = $ _____ and compare to the turns ratio determined from $a = V_p/N_s = 120$ V/12.6 V = 9.52. List any reasons for a difference in their levels.

Part 2   Voltage, Current, and Turns Ratio of the Transformer

(a) Construct the circuit of Fig. 15.3. Insert the measured resistor values. Set the input voltage to 12 V (*p-p*) using the oscilloscope.

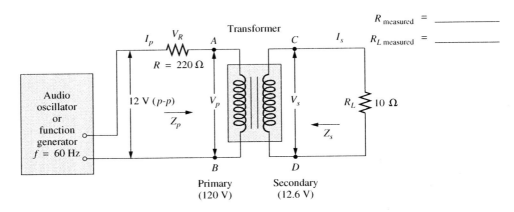

FIG. 15.3

Measure and record the primary and secondary rms voltages and V_R with the DMM and record in Table 15.1.

TABLE 15.1

V_p	
V_s	
V_R	
$a = N_p/N_s$	
I_s	
I_p (Eq. 15.3)	
I_p (V_R/R)	

Using the measured values above and Eq. (15.2), calculate the turns ratio N_p/N_s and record in Table 15.1.

Calculation:

Is this a step-up or step-down transformer? Explain why.

(b) Determine the secondary current I_s shown in Fig. 15.3 using Ohm's law ($I_s = V_s/R_L$) and record in Table 15.1. Then, using Eq. (15.3), calculate $I_{p(rms)}$ and record in Table 15.1.

Calculation:

(c) Determine the current $I_{p(rms)}$ shown in Fig. 15.3 using Ohm's law ($I_p = V_R/R$) and record in Table 15.1.

Calculation:

How does this measured value compare to the calculated value of part 2(b)?

(d) Set up the network of Fig. 15.4. Note the new position of terminals A through D. In essence, the transformer is hooked up in the reverse manner—that is, the load is hooked up to the rated 120-V terminals. Insert the measured resistor values. Set the input voltage to 4 V (p-p) using the oscilloscope.

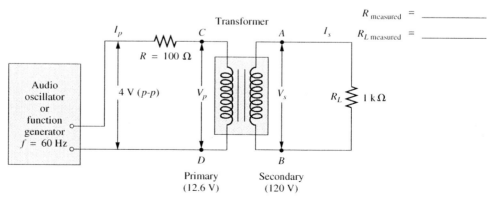

FIG. 15.4

Now measure the rms values of V_p, V_s, and V_R with the DMM and record in Table 15.2.

TABLE 15.2

V_p	
V_s	
I_s	
I_p	
$a = N_p/N_s$ (Eq. 15.2)	
$a = N_p/N_s$ (Eq. 15.3)	

Determine I_s from $I_s = V_s/R_L$ and I_p from $I_p = V_R/R$ using measured resistor values and record in Table 15.2.

Using Eq. (15.2), calculate the turns ratio and record in Table 15.2.

Calculation:

Using Eq. (15.3), calculate the turns ratio and record in Table 15.2. How does it compare to the above value?

Calculation:

Is the transformer being used as a step-up or step-down transformer? Explain why.

Part 3   Power Dissipation

Using the measured values of I_p, V_p, I_s, and V_s, calculate the apparent power ($P = VI$) of the primary and secondary circuits of the transformer of Fig. 15.3 and record in Table 15.3.

Calculation:

TABLE 15.3

$P_{primary}$	
$P_{secondary}$	

What conclusion can you draw from your calculations?

Part 4   Impedance Transformation

Reconnect the transformer as shown in Fig. 15.3.

(a) Using Eq. (15.4), calculate the reflected impedance (Z_p) for the network of Fig. 15.3 using the transformation ratio of part 2(a). Do not include R in Z_p, but use the measured value for R_L. Record the result in the top row of Table 15.4.

Calculation:

TABLE 15.4

	Calculated	Using V_p/I_p
Z_p (R_L only)		
Z_p (with 100 μF)		

(b) Calculate Z_p from the data of parts 2(a) and (c) in Table 15.1; that is, use $Z_p = V_p/I_p$ and record the result in the second column of Table 15.4.

Calculation:

How does this value of Z_p compare to the calculated value in part 4(a)?

(c) Disconnect the circuit (Fig. 15.3) from the oscillator or generator. Connect the 100-μF capacitor in series with the 10-Ω resistor. Calculate the magnitude of the reflected impedance Z_p for this circuit using the measured value for R_L and record in the first column of the second row.

Calculation:

(d) Connect the circuit to the oscillator or generator. Measure I_p and V_p and calculate the magnitude of Z_p. Record the result in Table 15.4.

Calculation:

How does this value of Z_p compare to the value calculated in part 4(c)?

EXERCISES

1. For the configuration of Fig. 15.5, determine each value.

    (a) V_s

    (b) I_s

    (c) I_p

    (d) Z_p

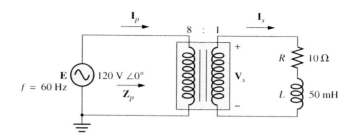

FIG. 15.5

$V_s =$ _____ , $I_s =$ _____

$I_p =$ _____ , $Z_p =$ _____

2. For the configuration of Fig. 15.6, determine each value.

  (a) Z_p

  (b) a

  (c) V_s

  (d) I_s

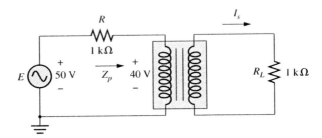

FIG. 15.6

$Z_p =$ _____ , $a =$ _____

$V_s =$ _____ , $I_s =$ _____

3. Using PSpice or Multisim, find the secondary voltage and current for the network of Fig. 15.3. Compare with the results of part 2 and attach any pertinent printouts.

EXPERIMENT ac **16**

Pulse Waveforms

OBJECTIVES

1. Become familiar with the terminology applied to pulse waveforms.
2. Note the effect of a dc level on a pulse waveform.
3. Determine the *R-C* response to a square-wave input.
4. Note the effect of the time constant on an *R-C* circuit on its pulse response.

EQUIPMENT REQUIRED

Resistors

1—1-kΩ

Capacitors

1—0.01-μF

Instruments

1—DMM
1—Oscilloscope
1—Square-wave or function generator
1—dc Power supply

EQUIPMENT ISSUED

TABLE 16.0

Item	Manufacturer and Model No.	Laboratory Serial No.
DMM		
Oscilloscope		
Generator		
dc Power supply		

RÉSUMÉ OF THEORY

Pulse Waveforms

Our analysis thus far has been limited to sinusoidal functions. In this experiment, we will examine pulse waveforms and the response of an R-C circuit to a square-wave input.

If the switch of Fig. 16.1(a) were opened and closed at a periodic rapid rate, the voltage across the resistor R would appear as shown in Fig. 16.1(b). As noted in Fig. 16.1(b), if the switch is open, the output is 0 V, since the current I is 0 A and $V_R = IR = (0)R = 0$ V. Once the switch is closed, V_R jumps to E volts. The difference between the lower and upper levels of the generated pulses is called the *step amplitude*. The abrupt change in voltage level is called an *electrical step*. As noted in Fig. 16.1(b), electrical steps can be defined as *positive-going* or *negative-going*. The pulses of Fig. 16.1(b) are ideal pulses since the sides are shown as absolutely vertical. Unfortunately, this is not the case in practical situations. A finite time (although often very small) is required for the voltage to change levels. A few important definitions associated with pulse waveforms are provided below.

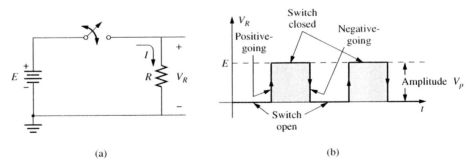

(a)   (b)

FIG. 16.1

Pulse Width (t_p):

The time between the two points on a pulse where the magnitude is one-half the maximum amplitude, as noted in Fig. 16.2.

Rise Time (t_r):

The time required for the voltage to go from $0.1V_p$ to $0.9V_p$, as noted in Fig. 16.2.

Fall Time (t_f):

The time required for the voltage to drop from 0.9 V_p to 0.1 V_p, as noted in Fig. 16.2.

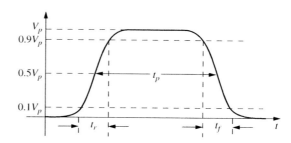

FIG. 16.2

Pulse Repetition Frequency (prf):

The rate at which the pulses of a periodic waveform repeat in a pulse train—actually the frequency of the periodic waveform. It is related to the period T by prf $= 1/T$. See Fig. 16.3.

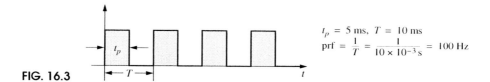

$$t_p = 5 \text{ ms}, \ T = 10 \text{ ms}$$
$$\text{prf} = \frac{1}{T} = \frac{1}{10 \times 10^{-3} \text{s}} = 100 \text{ Hz}$$

FIG. 16.3

Duty Cycle (D):

Since pulse waveforms often do not encompass 50% of the period, as shown in Fig. 16.4, the duty cycle, defined by $D = t_p/T$, is often applied to describe the pulse train.

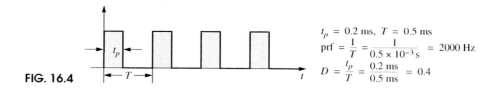

$$t_p = 0.2 \text{ ms}, \ T = 0.5 \text{ ms}$$
$$\text{prf} = \frac{1}{T} = \frac{1}{0.5 \times 10^{-3} \text{s}} = 2000 \text{ Hz}$$
$$D = \frac{t_p}{T} = \frac{0.2 \text{ ms}}{0.5 \text{ ms}} = 0.4$$

FIG. 16.4

A special case of the pulse waveform is the square wave appearing in Fig. 16.5, with an average value other than 0 V. The square wave has a duty cycle of 0.5 and an average value determined by $V_{av} = (D)(V_p) + (1 - D)(V_b)$, where the base-line value is as indicated in Fig. 16.5.

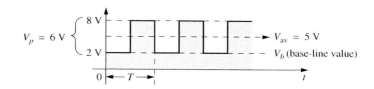

FIG. 16.5

For the waveform of Fig. 16.5,

$$V_{av} = (D)(V_p) + (1 - D)(V_b) \qquad\qquad (16.1)$$
$$= (0.5)(8\ V) + (1 - 0.5)(2\ V)$$
$$= 4\ V + 1V = 5\ V$$

as is confirmed by Fig. 16.5.

R-C Response to a Square Wave

Consider the R-C circuit of Fig. 16.6 with the square-wave input. Assume that the capacitor is initially uncharged and that the signal is applied at $t = 0$ s. The mathematical expression relating the voltage $V_C(t)$ to the other parameters of the circuit is the following:

$$V_C(t) = V_f + (V_i - V_f)e^{-t/\tau} \qquad\qquad (16.2)$$

where V_f is the final value of the voltage $V_C(t)$, V_i is the initial value of the voltage $V_C(t)$, and τ is the time constant of the circuit $= RC$.

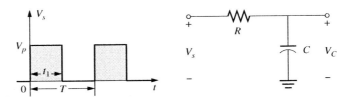

FIG. 16.6

For the conditions of Fig. 16.6, $V_f = V_p$ and $V_i = 0$ V. Substituting into Eq. (16.2),

$$V_C(t) = V_p + (0 - V_p)e^{-t/\tau} \qquad\qquad (16.3)$$

and

$$V_C(t) = V_p(1 - e^{-t/\tau}) \qquad\qquad (16.3)$$

If we choose a frequency such that $T = 2\tau$, then t_1 of Fig. 16.6 will be defined by $t_1 = \tau$ and the voltage $V_C(t)$ will be determined by the following:

$$V_C(t) = V_p(1 - e^{-\tau/\tau}) = V_p(1 - e^{-1}) = V_p(1 - 0.367)$$
$$= 0.633\ V_p$$

as shown in Fig. 16.7.

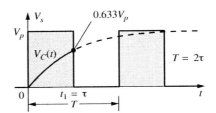

FIG. 16.7

If we reduce the frequency such that $T = 10\tau$ or $t_1 = 5\tau$, then $V_C(t)$ will for all practical purposes reach its final value, as indicated in Fig. 16.8.

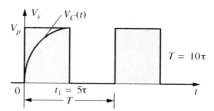

FIG. 16.8

For intermediate values of t_1 established by the applied frequency, the voltage $V_C(t)$ will reach the levels indicated in Table 16.1.

TABLE 16.1

t_1	$e^{-t/\tau}$	$(1-e^{-t/\tau})$	$V_C(t)$
τ	0.367	0.633	0.633 V_p
2τ	0.135	0.865	0.865 V_p
3τ	0.049	0.951	0.951 V_p
4τ	0.018	0.982	0.982 V_p
5τ	0.006	0.994	0.994 V_p

After the first positive pulse, the applied voltage abruptly drops to zero. The parameters of Eq. (16.2) then change to the following:

$$V_f = 0\text{ V} \quad \text{and} \quad V_i = \text{value of } V_C(t) \text{ at } t = t_1$$

And substituting into Eq. (16.2) yields

$$V_C(t) = V_f + (V_i - V_f)e^{-t/\tau} = 0 + (V_i - 0)e^{-t/\tau}$$

and

$$\boxed{V_C(t) = V_i e^{-t/RC}} \tag{16.4}$$

For the case of $t_1 = \tau$,

$$\begin{aligned} V_C(t) = V_i e^{-t/RC} &= (0.633V_p)e^{-\tau/\tau} \\ &= (0.633V_p)e^{-1} \\ &= (0.633V_p)(0.367) \\ &= 0.232V_p \end{aligned}$$

as shown in Fig. 16.9.

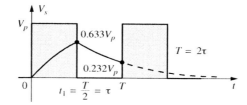

FIG. 16.9

For the second pulse, the new initial value is 0.232 V_p with a V_f equal to V_p. In time, the waveform will settle down to that appearing in Fig. 16.10 for $t_1 = T/2 = \tau$.

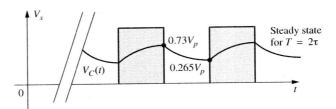

FIG. 16.10

For the case of $T = 10\tau$, the capacitor will discharge to zero and V_i will be 0 V for the second pulse, resulting in the same waveform as Fig. 16.8.

For the other possibilities of Table 16.2, the resulting waveform will be like the one in Fig. 16.10 but with different transition values.

PROCEDURE

Part 1   Repetition Rate and Duty Cycle

(a) Set the function generator to the square-wave mode and set the frequency to 1 kHz. Connect the generator to the oscilloscope as shown in Fig. 16.11.

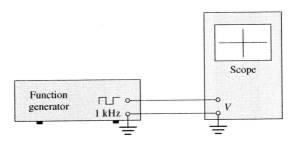

FIG. 16.11

Adjust the amplitude control until a 6-V (p-p) signal is obtained. Then adjust the sweep control until *two* full periods appear on the screen. Using the scope display, adjust the fine-control on the frequency of the function generator to ensure that the output frequency is exactly 1 kHz. That is, using the horizontal sensitivity, determine how many horizontal divisions should encompass a 1-kHz signal and make the necessary adjustments. Using the resulting display, complete Table 16.2.

TABLE 16.2

Pulse width	
Period	
Repetition rate	
Duty cycle	

Move the vertical input mode control from ac to dc and back again. Is there any change in the appearance or location of the display? Is there any average dc present?

(b) Place the mode switch to ground (GND) and set the zero reference line in the center of the screen. Return the switch to the dc position and note that the square wave has an equal positive and negative pulse.

Determine the average value using Eq. (16.1).

Calculation:

(c) Leave the generator and scope settings unchanged and construct the network of Fig. 16.12. The dc supply should be left "floating." That is, do not connect the ground terminal of the dc supply to any point in the network. Set the dc supply to 3 V and adjust the generator for a 6-V (*p-p*) signal at 1 kHz. Then set the zero reference line at the bottom of the screen and note the effect of switching the mode control of the oscilloscope from ac to dc and back.

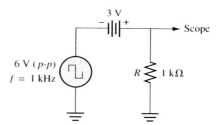

FIG. 16.12

With the mode control of the oscilloscope in the dc position, sketch the waveform below and indicate the zero reference line.

(d) Using Eq. (16.1), determine the average value of the waveform of (c).

Calculation:

Part 2   R-C Response to a Square-Wave Input for $T = 100\tau$ ($\tau = T/100$ or $5\tau = T/20$)

(a) Construct the network of Fig. 16.13. Insert the measured value of the resistor R.

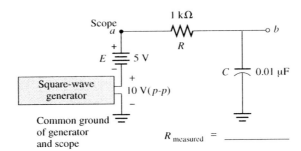

FIG. 16.13

$R_{measured} = $ _____

(b) Set the dc power supply (ungrounded) to 0 V, set the square-wave generator to 1 kHz, and connect the scope to point a. Adjust the amplitude of the generator to 10 V (p-p). Using the mode switch, establish the bottom line of the screen as the zero reference level. With the mode switch on dc, adjust the power supply until the bottom of the negative half-cycle rests on the zero reference line. In addition, adjust the frequency of the generator and the horizontal sensitivity control until there are only two cycles on the screen. Draw the resulting waveform for one period of the applied square wave on Graph 16.1.

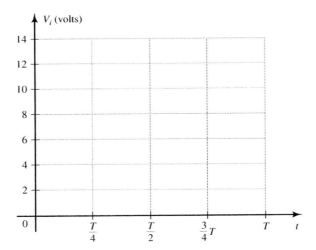

GRAPH 16.1

(c) Determine the time constant τ of the network and record in Table 16.3. Use the measured value of R.

Calculation:

(d) Determine T and $T/2$ of the applied square wave and record in Table 16.3. Compare to the time constant calculated in (c).

(e) Calculate the value of $V_C(t)$ at $t = t_1 = T/2$ using Eq. (16.3) and record in Table 16.3.

Calculation:

TABLE 16.3

τ	
T	
$T/2$	
$V_C(T/2)$	
$V_C(T/4)$	
$V_C(T/4)$	

(f) What does the value of part 2(e) indicate to you?

(g) Place the scope at point *b* of Fig. 16.13 and draw the trace on Graph 16.2 .

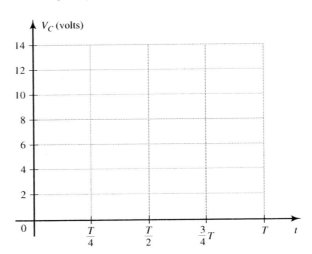

GRAPH 16.2

(h) Switch the positions of the resistor and capacitor and sketch the waveform across *R* on Graph 16.3. Make the adjustments necessary to ensure that the total waveform is on the screen.

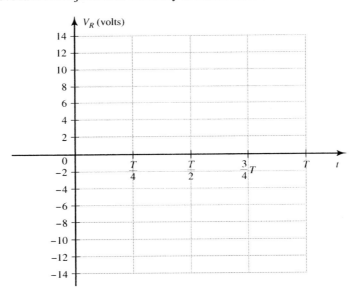

GRAPH 16.3

(i) At $t = T/4$, does $V_s = V_R + V_C$? Use the waveforms sketched above to determine V_R and V_C at $t = T/4 = 25\tau$ and record in Table 16.3.

Calculation:

Part 3   *R-C* Response for $T = 10\tau(\tau = T/10$ or $5\tau = T/2)$

(**a**) Reconstruct the network of Fig. 16.13 and connect the scope to point a. This part of the experiment will examine the response for $T = 10\tau$ or $T/2 = 5\tau$. The required frequency can be determined from the period T. Calculate T, $T/2$, and the frequency and record in Table 16.4.

Calculation:

TABLE 16.4

T	
$T/2$	
f	
$V_C(T/2)$	
$V_C(T/4)$	
$V_R(T/4)$	

(**b**) Set the input to the same signal as in part 2(b) but at the frequency calculated in part 3(a). Calculate the magnitude of $V_C(t)$ at $t = t_1 = T/2 = 5\tau$ and record in Table 16.4.

Calculation:

(**c**) Connect the scope to point b and sketch the waveform on the same graph used for part 2(g). Is $V_C(t)$ the value forecasted in part 3(b)?

(**d**) Switch the positions of the capacitor and resistor and sketch V_R on Graph 16.3.

(**e**) At $t = T/4$, does $V_s = V_R + V_C$? Use the waveforms sketched above to determine V_R and V_C at $t = T/4 = 2.5\tau$ and record in Table 16.4.

Calculation:

Part 4   R-C Response for $T = 2\tau (\tau = T/2$ or $5\tau = 2.5T)$

(a) Repeat part 3(a) for $T = 2\tau$ or $T/2 = \tau$. Record the resulting T, $T/2$, and f in Table 16.5.

Calculation:

TABLE 16.5

T	
$T/2$	
f	
$V_C(T/2)$	
$V_C(T)$	
$V_C(^3/_2T)$	
$V_C(2T)$	
$V_C(2.5T)$	

(b) Set the input to the same signal as in part 2(b) but at the frequency calculated in part 4(a). Determine the magnitude of $V_C(t)$ at $t = t_1 = T/2 = \tau$ and record in Table 16.5.

Calculation:

Plot the curve of V_C from $t = 0$ s to $t = T/2$ on Graph 16.4.

(c) Using the V_C of part 4(b) as the initial value for the discharge cycle between $T/2$ and T, determine the value of V_C at $t = T$ and record in Table 16.5.

Calculation:

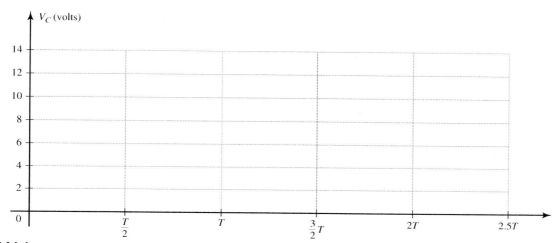

GRAPH 16.4

Plot the curve of V_C from $t = T/2$ to T on Graph 16.4.

Using the value of V_C at $t = T$ as the initial value for the charging cycle between T and $\frac{3}{2}T$, calculate the value of V_C at $t = \frac{3}{2}T$ and record in Table 16.5.

Calculation:

Plot the curve of V_C from $t = T$ to $\frac{3}{2}T$ on Graph 16.4.

Using the value of V_C at $\frac{3}{2}T$ as the initial value for the discharge cycle between $\frac{3}{2}T$ and $2T$, calculate the value of V_C at $t = 2T$ and record in Table 16.5.

Calculation:

Plot the curve of V_C from $\frac{3}{2}T$ to $2T$ on Graph 16.4.

And, finally, using the value of V_C at $2T$ as the initial value of the charging cycle between $2T$ and $2.5T$, calculate the value of V_C at $t = 2.5T$ and record in Table 16.5.

Calculation:

Plot the curve of V_C from $2T$ to $2.5T$ on Graph 16.4.

(d) Connect the oscilloscope to point b and sketch the resulting waveform on Graph 16.4. Be sure to label the calculated and measured curves. How do the two curves compare? Where are they different? Why?

(e) Switch the positions of the capacitor and resistor and sketch V_R (from calculated values) on Graph 16.5 using the same sequence of steps provided above for V_C. You may also use the fact that at any instant of time $V_s = V_C + V_R$ to plot your curve.

(f) Apply the oscilloscope and record the waveform for V_R on Graph 16.5. Be sure to label the calculated and measured curves. How do the two curves compare? Where are they different? Why?

(g) What is the effect of a further increase in frequency on the voltage across the capacitor or resistor? Explain why the waveform changes as it does.

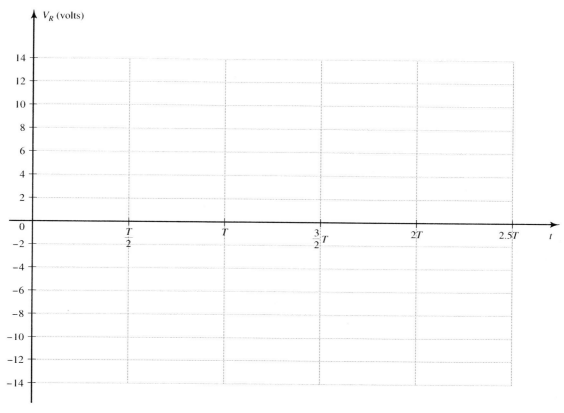

GRAPH 16.5

EXERCISES

1. For the input v_s appearing in Fig. 16.14, sketch the waveform for v_o.

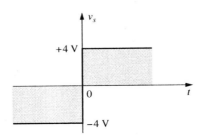

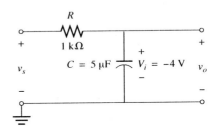

FIG. 16.14

2. Use the input v_s appearing in Fig. 16.15.

   (a) Sketch the waveform for v_o.

   (b) Change C to 0.1 μF and sketch the waveform for v_o.

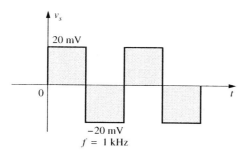

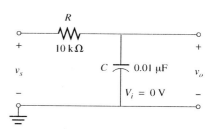

FIG. 16.15

Name:_____

Date:_____

Course and Section:_____

Instructor:_____

Currents and Voltages in Balanced Three-Phase Systems

OBJECTIVES

1. Measure the line and phase voltages and currents of a balanced Δ, Y, and Y-Δ configuration.
2. Verify the basic equations between line and phase voltages and currents for three-phase balanced systems.
3. Demonstrate the use of a phase-sequence indicator.

EQUIPMENT REQUIRED

Resistors

3—100-Ω, 330-Ω, 18-kΩ, 33-kΩ (2-W)

Instruments

1—DMM

2—Phase-sequence indicator (if available)

Power Supply

Three-phase, 120/208-V, 60-Hz

EQUIPMENT ISSUED

TABLE 17.0

Item	Manufacturer and Model No.	Laboratory Serial No.
DMM		
Phase-sequence indicator		

RÉSUMÉ OF THEORY

In a balanced three-phase system, the three-phase voltages are equal in magnitude and 120° out of phase. The same holds true for the line voltages, phase currents, and line currents. The system is balanced if the impedance of each phase is the same. For the balanced Y load, the line current is equal to the phase current, and the line voltage is $\sqrt{3}$ (or 1.73) times the phase voltage. For the balanced Δ load, the line and phase voltages are equal, but the line current is $\sqrt{3}$ (or 1.73) times the phase current.

The relationship used to convert a balanced Δ load to a balanced Y load is

$$\boxed{Z_Y = \frac{Z_\Delta}{3}} \tag{17.1}$$

Since the sum (phasor) of three equal quantities 120° out of phase is zero, if a neutral wire is connected to a balanced three-phase, Y-connected load, the current in this neutral wire will be zero. If the loads are unbalanced, the current will not be zero.

The phase sequence of voltage does not affect the magnitude of currents in a balanced system. However, if the load is unbalanced, changing the phase sequence will change the magnitude of the currents. The resulting sequence of voltages is determined by the direction of rotation of a three-phase motor.

PROCEDURE

Note: **Be absolutely sure that the power is turned off when making changes in the network configuration or inserting and removing instruments.**

Part 1 Resistors in a Balanced Δ

Construct the network of Fig. 17.1. Insert the measured resistor values.

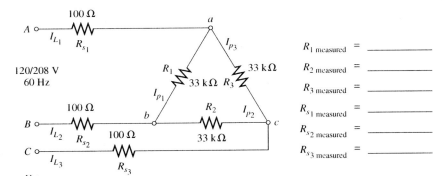

FIG. 17.1

(a) Measure the three-phase voltages and line voltages and record in Table 17.1.

TABLE 17.1

Line voltages	V_{AB}	V_{BC}	V_{CA}
Phase voltages	V_{ab}	V_{bc}	V_{ca}

What is the relationship between the line voltage and the phase voltage in a balanced Δ-connected load? Ignore the effects of the sensing resistors R_{S_1}, R_{S_2}, and R_{S_3}.

(b) Measure the voltage across the sensing resistors with the DMM and record in Table 17.2.

Calculate the line currents using the sensing voltages, Ohm's law, and measured resistor values. Show all work! Record the results in Table 17.2.

Calculation:

TABLE 17.2

$V_{R_{s_1}}$	$V_{R_{s_2}}$	$V_{R_{s_3}}$
$I_{R_{s_1}}$	$I_{R_{s_2}}$	$I_{R_{s_3}}$
I_{L_1}	I_{L_2}	I_{L_3}
I_{p_1}	I_{p_2}	I_{p_3}
I_{L_1}/I_{p_1}	I_{L_2}/I_{p_2}	I_{L_3}/I_{p_3}

(c) Using Ohm's law, the measurements of part 2(a), and measured resistor values, calculate the phase currents. Show all work! Record the results in Table 17.2.

Calculation:

(d) For the balanced Δ-connected load, how are the line currents related (in magnitude)?

(e) For the balanced Δ-connected load, how are the phase currents related (in magnitude)?

(f) Determine the ratios I_{L_1}/I_{P_1}, I_{L_2}/I_{P_2}, I_{L_3}/I_{P_3}. How close are the ratios to the $\sqrt{3} = 1.732$ factor described in the Résumé of Theory? Record the results in Table 17.2.

Calculation:

Part 2 Resistors in a Balanced Y

Construct the network of Fig. 17.2. Insert the measured resistor values.

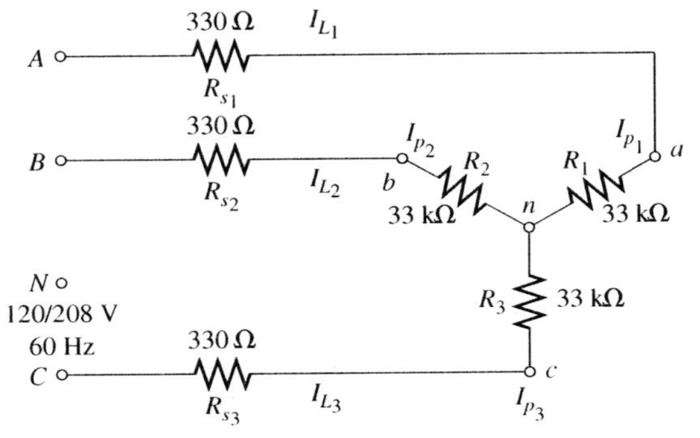

R_1 measured = ________

R_2 measured = ________

R_3 measured = ________

R_{s1} measured = ________

R_{s2} measured = ________

R_{s3} measured = ________

FIG. 17.2

(a) Measure the three-phase voltages and line voltages and record in Table 17.3.

TABLE 17.3

Line voltages	V_{AB}	V_{BC}	V_{CA}
Phase voltages	V_{an}	V_{bn}	V_{cn}

What is the relationship between the magnitudes of the line voltages and the phase voltages of a balanced Y-connected load? Ignore the effects of the sensing resistors R_{S_1}, R_{S_2}, and R_{S_3}.

(b) Measure the voltage across the sensing resistors with the DMM and record in Table 17.4.

Calculate the line currents using the sensing voltages, Ohm's law, and measured resistor values. Show all work. Record the results in Table 17.4.

Calculation:

TABLE 17.4

$V_{R_{S_1}}$	$V_{R_{S_2}}$	$V_{R_{S_3}}$
$I_{R_{S_1}}$	$I_{R_{S_2}}$	$I_{R_{S_3}}$
I_{L_1}	I_{L_2}	I_{L_3}
I_{p_1}	I_{p_2}	I_{p_3}

(c) Using Ohm's law, the measurements of part 3(a), and measured resistor values, calculate the phase currents. Show all work! Record the results in Table 17.4.

Calculation:

(d) For the balanced Y-connected load, how are the line currents related (in magnitude)?

(e) For the balanced Y-connected load, how are the phase currents related (in magnitude)?

(f) For the balanced Y-connected load, how are the line and phase currents related?

Part 3   Resistors in a Y-Δ Configuration

(a) Construct the network of Fig. 17.3. For this part of the experiment, use the nominal (color-coded) values of the resistors.

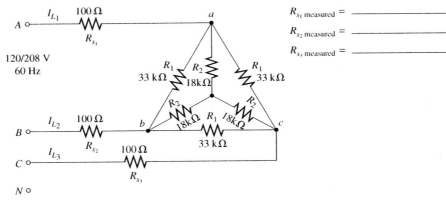

FIG. 17.3

(b) Read the line voltages V_{ab}, V_{bc}, and V_{ca} and the voltages V_{R_1}, V_{R_2} and V_{R_3} and calculate the line currents. Record all the results in Table 17.5.

Calculation:

TABLE 17.5

Line voltages	V_{ab}	V_{bc}	V_{ca}
Sensing voltages	$V_{R_{s_1}}$	$V_{R_{s_2}}$	$V_{R_{s_3}}$
Line currents	I_{L_1}	I_{L_2}	I_{L_3}

(c) Convert the Y to a Δ using the nominal resistor values and combine the resulting Δ with the original Δ to form a single Δ configuration. Sketch the resulting Δ below and indicate the magnitude of each resistor of the configuration.

(d) Calculate the phase current for each branch of the Δ, and then calculate the line currents. Record the average value of the phase and line currents in Table 17.6.

Calculation:

TABLE 17.6

	Average I_p	Average I_L
Part 3(d)		
Part 3(g)		

(e) How does the calculated level of line current in part 3(d) compare with the measured value of part 3(b)?

(f) Now convert the Δ to a Y using the nominal resistor values and combine the resulting Y with the original Y to form a single Y configuration. Sketch the resulting Y here and indicate the magnitude of each resistor of the configuration.

(g) Calculate the phase current for each branch of the Y, and then calculate the line currents. Record the average value of the phase and line currents in Table 17.6.

Calculation:

(h) How does the calculated level of line current in part 3(g) compare with the measured value of part 3(b)?

(i) Based on these results, have the Y-Δ and Δ-Y conversion equations been verified?

Part 4   Phase-Sequence Indicator (If Available)

Connect the sequence indicator to the power source as shown in Fig. 17.4. Turn on the power. If the 1-2-3 light is brighter than the 3-2-1 light, then the phase sequence is 1-2-3 or $\mathbf{E}_{12}(\mathbf{E}_{AB})$, $\mathbf{E}_{23}(\mathbf{E}_{BC})$, and $\mathbf{E}_{31}(\mathbf{E}_{CA})$, which means that if $\mathbf{E}_{12}$ is the reference, then $\mathbf{E}_{23}$ lags $\mathbf{E}_{12}$ by 120°, and $\mathbf{E}_{31}$ lags $\mathbf{E}_{23}$ by 120°. If the 3-2-1 lamp is brighter, then the sequence is 3-2-1. Record the phase sequence of the line voltages.

Sequence = _____

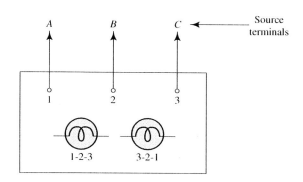

FIG. 17.4

Read the magnitude of the line voltages (E_{AB}, E_{BC}, and E_{CA}) and the phase voltages (E_{AN}, E_{BN}, and E_{CN}) using the DMM and record in Table 17.7.

TABLE 17.7

Line voltages	E_{AB}	E_{BC}	E_{CA}
Phase voltages	E_{AN}	E_{BN}	E_{CN}

Draw the phasor diagram of the line voltages showing magnitude and phase angle if E_{AB} is the reference voltage at an angle of $0°$.

EXERCISES

1. (a) If a phase sequence indicator indicates a 1-2-3 (*A-B-C*) sequence for the line voltages, write the line voltages for your supply in phasor form. Use $\mathbf{E}_{AB} = E_{AB} \angle 0°$.

$\mathbf{E}_{AB} = $ _____, $\mathbf{E}_{BC} = $ _____, $\mathbf{E}_{CA} = $ _____

(b) Then determine the phase currents for the load of Fig. 17.1 in phasor form.

$\mathbf{I}_{p_1} = $ _____, $\mathbf{I}_{p_2} = $ _____, $\mathbf{I}_{p_3} = $ _____

(c) Finally, sketch the phasor diagram of the phase currents.

2. (a) If the phase sequence indicator shows a 3-2-1 (*C-B-A*) sequence for the line voltages, write the line voltages for your supply in phasor form. Use $\mathbf{E}_{AB} = \mathbf{E}_{AB} \angle 0°$.

$\mathbf{E}_{AB} =$ _____ , $\mathbf{E}_{BC} =$, _____ $\mathbf{E}_{CA} =$ _____

(b) Then determine the phase currents for the load of Fig. 17.2 in phasor form.

$\mathbf{I}_{p_1} =$ _____ , $\mathbf{I}_{p_2} =$ _____ , $\mathbf{I}_{p_3} =$ _____

(c) Finally, determine the line currents for the load of Fig. 17.2 in phasor form and sketch the phasor diagram for the line currents.

Name: _____

Date: _____

Course and Section: _____

Instructor: _____

Power Measurements in Three-Phase Systems

OBJECTIVES

1. Determine the power delivered to a three-phase load using three wattmeters.
2. Review the procedure for determining the power to a three-phase load using two wattmeters.
3. Determine the phase angle of the load from the wattmeter readings.

EQUIPMENT REQUIRED

Resistors

3—250-Ω (200-W or more), wire-wound

Instruments

1—DMM

2—1000-W-maximum wattmeters (HPF)

Power Supply

Three-phase, 120/208-V, 60-Hz

EQUIPMENT ISSUED

TABLE 18.0

Item	Manufacturer and Model No.	Laboratory Serial No.
DMM		
Wattmeter		
Wattmeter		

RÉSUMÉ OF THEORY

The total average power to a three-phase load circuit is given by Eqs. (18.1) and (18.2):

$$\boxed{P_T = 3E_p I_p \cos \theta}$$ (18.1)

where E_p is the phase voltage, I_p the phase current, $\cos \theta$ the power factor, and θ the phase angle between E_p and I_p.

$$\boxed{P_T = \sqrt{3} E_L I_L \cos \theta}$$ (18.2)

where E_L is the line voltage, I_L is the line current, $\cos \theta$ is the power factor, and θ is the phase angle between E_p and I_p.

The power to a three-phase load circuit can be measured in two ways:

1. Three-wattmeter method
2. Two-wattmeter method

THREE-WATTMETER METHOD

The connections for the wattmeters for a Δ and a Y circuit are shown in Fig. 18.1. The total power P_T is equal to the sum of the wattmeter readings.

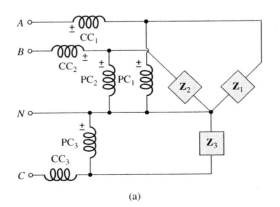

(a)

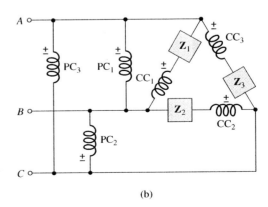

(b)

FIG. 18.1

Note that in each case, connections must be made to the individual load in each phase, which is sometimes awkward, if not impossible. A more feasible and economical method is the two-wattmeter method.

TWO-WATTMETER METHOD

With the two-wattmeter method, it is not necessary to be concerned with the accessibility of the load. The two wattmeters are connected as shown in Fig. 18.2.

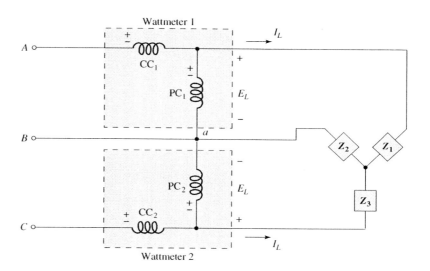

FIG. 18.2

It can be shown that the two-wattmeter method of measuring total average power yields the correct value under any conditions of load.

If we plot a curve of power factor versus the ratio of the wattmeter readings P_1/P_2, where P_1 is always the smaller reading, we obtain the curve appearing in Fig. 18.3.

The graph shows that if the power factor is equal to one (resistive circuit), then $P_1 = P_2$ and the readings are additive. If the power factor is zero (reactive), then $P_1 = P_2$; but P_1 is negative, so $P_T = 0$. At a 0.5 power factor, one wattmeter reads zero, and the other reads the total power. It is indicative from the graph that if $F_p > 0.5$, the readings are to be added; and if $F_p < 0.5$, the readings are to be subtracted.

The two-wattmeter method will read the total power in a three-wire system even if the system is unbalanced. For three-phase, four-wire systems (balanced or unbalanced), three wattmeters are required.

It can be shown that the power read by each wattmeter is given by the following equations:

$$\boxed{P_2 = E_L I_L \cos(\theta - 30°)} \tag{18.3}$$

$$\boxed{P_1 = E_L I_L \cos(\theta + 30°)} \tag{18.4}$$

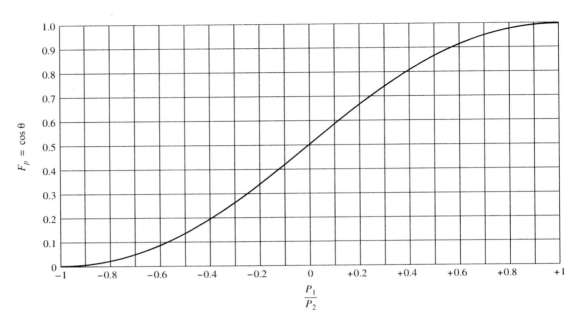

FIG. 18.3

$$P_T = P_2 \pm P_1$$ (the algebraic sum of P_1 and P_2, where P_1 is the smaller reading)   **(18.5)**

Another advantage of the two-wattmeter method is that it allows us to calculate the phase angle (θ) and, therefore, the power factor (F_p) using the expressions

$$\tan \theta = \sqrt{3}\, \frac{P_2 - P_1}{P_2 + P_1}$$ **(18.6)**

$$F_p = \cos \theta$$ **(18.7)**

PROCEDURE FOR CONNECTING THE WATTMETER USING THE TWO-WATTMETER METHOD

Connect the potential coils (PC) in the circuit last, and use the insulated alligator clip leads for their connections to line *B*. (See Fig. 18.2.) Close the breaker. If either wattmeter reads down-scale, shut down and reverse the current coil (CC) of the wattmeter that reads down-scale. Apply power again and read both wattmeters.

To find the sign of the lower-reading wattmeter, proceed as follows. If wattmeter 1 is the lower reading, unclip lead *a* of the PC coil and touch to any point on line *A*. If the wattmeter deflects up-scale, the reading is to be added (+). If the deflection is down-scale, the reading is negative and is to be subtracted (−). If wattmeter 2 is the lower-reading wattmeter, then the PC lead *a* is touched to line *C* for the test.

PROCEDURE

Part 1   Balanced Δ-Connected Resistor Loads

(**a**) Insert the measured values of R_1, R_2, and R_3 in Fig. 18.4.

(**b**) Construct the network of Fig. 18.4. Refer to the procedure for connecting the wattmeter. *Exercise caution at all times!* If in doubt, ask your instructor.

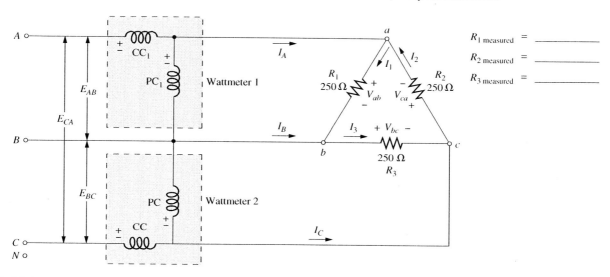

FIG. 18.4

Measure the magnitude of the voltages E_{ab}, E_{bc}, and E_{ca} using the DMM and record in Table 18.1.

Are the line voltages the same as the phase voltages of the load V_{ab}, V_{bc}, and V_{ca}? Explain.

TABLE 18.1

Line voltages	E_{cb}	E_{bc}	E_{ca}
Phase currents	I_1	I_2	I_3
Line currents	I_A	I_B	I_C
Power	P_{R_1}	P_{R_2}	P_{R_3}

(c) Calculate the magnitude of the phase currents I_1, I_2, and I_3 using measured resistor values and record in Table 18.1.

Calculation:

Calculate the magnitude of the line currents I_A, I_B, and I_C and record in Table 18.1. Show all work!

Calculation:

(d) Calculate the power dissipated by each resistor using the basic power equation $P = I^2R$ and record in Table 18.1.

Calculation:

Determine the total power using the above results and record in Table 18.2. Show all work.

Calculation:

TABLE 18.2

	Sum (Table 18.1)	Eq. (18.1)	Eq. (18.2)
P_T			

Determine the total power delivered using Eq. (18.1) and record in Table 18.2. Compare with the above results. Use the average value of the phase currents in the equation.

Calculation:

Determine the total power delivered using Eq. (18.2) and record in Table 18.2. Compare with the previous results. Use the average value of the line currents in the equation.

Calculation:

(e) Record the readings of the wattmeters in Table 18.3.

Using the above measurements, calculate the total power and record in Table 18.3.

Calculation:

TABLE 18.3

P_1	
P_2	
P_T	
θ	
F_p	
P_1/P_2	
F_P (Fig. 18.3)	

How does the measured value compare to the calculated values?

Determine the phase angle and power factor of the load and record in Table 18.3.

Calculation:

Compute the ratio P_1/P_2 using measured values and calculate the power factor from Fig. 18.3 and record in Table 18.3. How does it compare to the power factor obtained above?

(f) Work with another squad so that the total power can now be measured using three wattmeters. Hook up the wattmeters as shown in Fig. 18.1(b). Note that each potential coil is sensing the voltage level of one phase while the current coil is sensing the current level for that phase.

TABLE 18.4

P_1	
P_2	
P_3	
P_T	

Have your instructor check the connections before turning on the power. Record the power levels obtained in Table 18.4.

Using the above measurements, calculate the total power to the load and record in Table 18.4.

Calculation:

How does this power level compare to the total power obtained in part 1(e) (Table 18.3)?

How do the three power level readings compare? Was this expected? Why?

Part 2   Balanced Four-Wire, Y-Connected Resistor Load

(a) Construct the circuit of Fig. 18.5. Refer to the procedure for applying the two-wattmeter method. *Exercise caution at all times!* If in doubt, ask your instructor. Insert the measured value of each resistor.

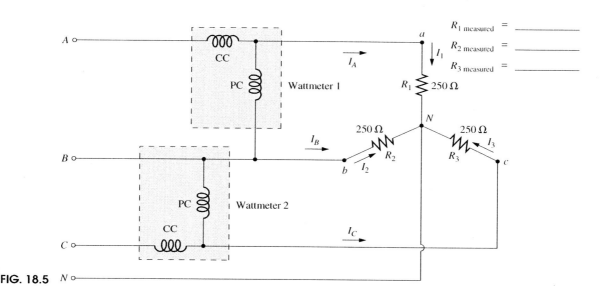

FIG. 18.5

Measure the magnitude of the phase voltages V_{an}, V_{bn}, and V_{cn} and record in Table 18.5.
Measure the live voltages E_{ab}, E_{bc}, and E_{ca} and record in Table 18.5.

TABLE 18.5

Phase voltages	V_{an}	V_{bn}	V_{cn}
Line voltages	E_{ab}	E_{bc}	E_{ca}
Ratios	E_{ab}/V_{an}	E_{bc}/V_{bn}	E_{ca}/V_{cn}
Phase currents	I_1	I_2	I_3
Line currents	I_A	I_B	I_C
Power	P_{R_1}	P_{R_2}	P_{R_3}

What is the relationship between (V_{an}, V_{bn}, and V_{cn}) and (E_{ab}, E_{bc}, and E_{ca})?

Calculate the ratios E_{ab}/V_{an}, E_{bc}/V_{bn}, and E_{ca}/V_{cn} and record in Table 18.5.
Do these ratios confirm the expected relationship?

(b) Calculate the magnitude of the phase currents I_1, I_2, and I_3 using measured resistor values and insert in Table 18.5.

Calculation:

How are these currents related to the line currents I_A, I_B, and I_C ? List the line currents in Table 18.5.

(c) Calculate the power delivered to each resistive element of the load and record in Table 18.5.

Calculation:

Determine the total power using the above results and record in Table 18.6. Show all work.

Calculation:

TABLE 18.6

	Sum (Table 18.5)	Eq. (18.1)	Eq. (18.2)
P_T			

Determine the total power delivered using Eq. (18.1) and record in Table 18.6. Compare with the above result. Use the average value of the phase currents in the equation.

Calculation:

Determine the total power delivered using Eq. (18.2) and record in Table 18.6. Compare with the above results. Use the average value of the line currents in the equation.

Calculation:

(d) Record the readings of the wattmeters in Table 18.7.

Using the above measurements, calculate the total power and record in Table 18.7.

Calculation:

TABLE 18.7

P_1	
P_2	
P_T	
θ	
F_p	
P_1/P_2	
F_p (Fig. 18.3)	

How does the measured value compare to the calculated values?

Determine the phase angle and power factor of the load and insert in Table 18.7.

Calculation:

Compute the ratio P_1/P_2 and calculate the power factor from Fig. 18.3 and insert in Table 18.7. How does it compare to the power factor obtained above?

Calculation:

(e) Work with another group so that the total power can now be measured using three wattmeters. Hook up the wattmeters as shown in Fig. 18.1(a). Note that each potential coil is sensing a phase voltage while the current coil is sensing the line current that equals the phase current. **Have your instructor check the connections before turning on the power.** Record the power levels in Table 18.8.

Using the above measurements, calculate the total power to the load and insert in Table 18.8.

TABLE 18.8

P_1	
P_2	
P_3	
P_T	

Calculation:

How does this power level compare to the total power obtained in part 1(d)?

How do the readings of the three wattmeters compare? Was this expected? Why?

EXERCISES

1. (a) Replace the resistor R_2 of Fig. 18.4 with a 100-Ω resistor and determine the reading of each wattmeter. Assume an A-B-C phase sequence and use nameplate values.

   (b) Calculate the ratio P_1/P_2 and determine F_p from Fig. 18.3.

   (c) Determine tan θ using Eq. (18.6) and then find F_p from Eq. (18.7). Compare the solutions of parts (b) and (c).

$P_1 = $ _____, $P_2 = $ _____, F_p (Fig. 18.3) = _____

F_p (Eq. (18.6)) = _____

2. (a) Replace the resistor R_3 of Fig. 18.4 with a series branch of a 100-Ω resistor and 100-Ω coil and determine the reading of each wattmeter. Assume an A-B-C phase sequence. R_1 and R_2 remain at 250 Ω.

   (b) Calculate the ratio P_1/P_2 and determine F_p from Fig. 18.3.

(c) Determine $\tan \theta$ using Eq. (18.6) and then find F_p from Eq. (18.6). Compare the solutions of parts (b) and (c).

$P_1 =$ _____, $P_2 =$ _____, F_p (Fig. 18.3) = _____

F_p (Eq. (18.6)) = _____

Parts List for the Experiments

dc (REQUIRED)

Resistors

1/4-W Film

    1—4.7 Ω, 6.8 Ω, 10 Ω, 47 Ω, 91 Ω, 100 Ω, 470 Ω, 680 Ω, 3.3 kΩ,
        4.7 kΩ, 6.8 kΩ, 22 kΩ, 47 kΩ, 470 kΩ, 1 MΩ

    2—220 Ω, 1.2 kΩ, 2.2 kΩ, 10 kΩ, 100 kΩ

    3—100 Ω, 330 Ω, 1 kΩ, 3.3 kΩ

    1—Unknown between 47 Ω and 220 Ω

1/2-W Film

    1—1 MΩ

1-W Film

    1—1 MΩ

2-W Film

    1—1 MΩ

Variable

    1—(0–1 kΩ) Potentiometer, linear, carbon

    1—(0–10 kΩ) Potentiometer, linear, carbon

    1—(0–250 kΩ) Potentiometer, linear, carbon

Capacitors

1—1 μF
2—100 μF (electrolytic)

Inductors

1—10 mH

Miscellaneous

1—Single-pole, single-throw switch
1—1-mA, 1000-Ω Meter movement
1—Neon glow lamp (rated 120 V)

Instruments

1—VOM
1—DMM
1—dc Power supply

dc (OPTIONAL)

Second dc power supply
Commercial Wheatstone bridge

ac (REQUIRED)

Resistors

1/4-W Film

1—4.7 Ω, 10 Ω, 22 Ω, 33 Ω, 47 Ω, 100 Ω, 220 Ω, 470 Ω, 1 kΩ, 1.2 kΩ, 2.2 kΩ, 3.3 kΩ, 6.8 kΩ

3—10 Ω, 100 Ω, 330 Ω, 10 kΩ, 18 kΩ, 33 kΩ
2-W (Any type—Experiment 17 only)

3—100 Ω, 330 Ω, 18 kΩ, 33 kΩ
200-W-wirewound (Experiment 18 only)

3—250 Ω
Variable

1—(0–1 kΩ) Potentiometer

1—(0–10 kΩ) Potentiometer

Capacitors

1—0.0047 μF, 0.001 μF, 0.01 μF, 0.1 μF, 0.2 μF, 0.47 μF, 1 μF, 15 μF
2—0.02 μF

Inductors

1—1 mH
2—10 mH

Miscellaneous

1—Transformer, 120 V primary, 12.6 V secondary
2—D Batteries and holders

Instruments

1—DMM
1—Audio oscillator or function generator
1—General purpose oscilloscope
1—dc Power supply
1—Frequency counter (if available)
1—Square-wave generator (Experiment 15 only)
2—1000 W HPF Watt meters (Experiment 17 only)

Power Sources

Three-phase 120/208-V supply (Experiments 16 and 17 only)

ac (OPTIONAL)

1—Phase sequence indicator

dc AND ac (TOTAL REQUIREMENTS)

Resistors

1/4-W Film

1—6.8 Ω, 10 Ω, 22 Ω, 33 Ω, 47 Ω, 91 Ω, 470 Ω, 680 Ω, 4.7 kΩ, 6.8 kΩ, 47 kΩ, 470 kΩ

2—220 Ω, 1.2 kΩ, 2.2 kΩ, 100 kΩ

3—100 Ω, 330 Ω, 1 kΩ, 3.3 kΩ, 10 kΩ, 18 kΩ, 33 kΩ

1—Unknown between 47 Ω and 220 Ω

1/2-W Film

1—1 MΩ

1-W Film

1—1 MΩ

2-W Film

1—1 MΩ

3—100 Ω, 330 Ω, 18 kΩ, 33 kΩ (Experiment 17 only)

200-W-wirewound

3—250 Ω (Experiment 18 only)

Variable

1—(0–1 kΩ) Potentiometer, linear, carbon

1—(0–10 kΩ) Potentiometer, linear, carbon

1—(0–250 kΩ) Potentiometer, linear, carbon

Capacitors

1—0.0047 μF, 0.001 μF, 0.01 μF, 0.1 μF, 0.2 μF, 0.47 μF, 1 μF, 15 μF
2—0.02 μF, 100 μF (electrolytic)

Inductors

1—1 mH
2—10 mH

Miscellaneous

1—Single-pole, single-throw switch
1—1 mA, 1000 Ω Meter movement
1—Transformer, 120 V primary, 12.6 V secondary
2—D Batteries and holders
1—Neon glow lamp (rated 120 V)

Instruments

1—VOM
1—DMM
1—dc Power supply
1—Audio oscillator or function generator
1—Frequency counter (if available)
1—General purpose oscilloscope
1—Square-wave generator (Experiment 15 ac only)
2—1000 W HPF Wattmeters (Experiment 17 ac only)

Power Sources

Three-phase 120/208-V supply (Experiments 17 ac and 18 ac only)

dc AND ac (OPTIONAL)

Second dc power supply
Commercial Wheatstone bridge
Phase sequence indicator

Resistor Color Coding

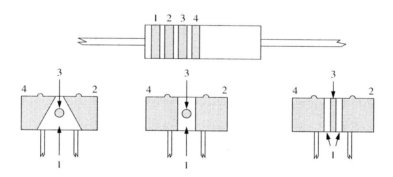

1 First significant digit
2 Second significant digit
3 Multiplying value
4 Tolerance (percent)

FIG. I.1

TABLE I.1 RETMA color code for resistors

Color	1 First Significant Digit	2 Second Significant Digit	3 Multiplier	4 Percent Tolerance
Silver			0.01	10
Gold			0.1	5
Black	(not used)	0	1.0	
Brown	1	1	10	1
Red	2	2	100	2
Orange	3	3	1000	3
Yellow	4	4	10,000	4
Green	5	5	100,000	
Blue	6	6	1,000,000	
Purple	7	7	10,000,000	
Gray	8	8	100,000,000	
White	9	9	1,000,000,000	
No color				20

EXAMPLE I.1 Application of resistor color code

Nominal Value	Percent Tolerance	Band				Tolerance	Range
		1	2	3	4		
22,000 Ω	± 20	Red	Red	Orange	No Band	$\pm 4400 \, \Omega$	17,600–26,400 Ω
100 Ω	± 5	Brown	Black	Brown	Gold	$\pm 5 \, \Omega$	95–105 Ω
10 Ω	± 10	Brown	Black	Black	Silver	$\pm 1 \, \Omega$	9–11 Ω
1.2 kΩ	± 5	Brown	Red	Red	Gold	$\pm 60 \, \Omega$	1140–1260 Ω
6.8 MΩ	± 20	Blue	Gray	Green	No Band	$\pm 1.36 \, M\Omega$	5.44–8.16 MΩ

APPENDIX II

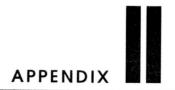

Capacitor Color Coding

MOLDED CERAMIC CAPACITORS

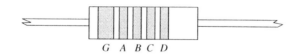

A First significant digit
B Second significant digit
C Multiplying value
D Tolerance
G Temperature coefficient

FIG. II.1

TABLE II.1   Color code for molded ceramic capacitors

Color	A First Significant Digit	B Second Significant Digit	C Multiplier	D Tolerance Value Greater Than 10 pF (%)	Value 10 pF or Less (pF)	G Temperature Coefficient (parts per million per °C)
Black	(not used)	0	1	20	2.0	0
Brown	1	1	10	1		−30
Red	1	1	100	2		−80
Orange	3	3	1000			−150
Yellow	4	4	10,000			−220
Green	5	5	100,000	5	0.5	−330
Blue	6	6	1,000,000			−470
Violet	7	7	10,000,000			−750
Gray	8	8	0.01		0.25	+30
White	9	9	0.1	10	1.0	+550

Resistance Measurements (VOM)

Never Make Resistance Measurements on a Live Circuit!

Select the two leads used for resistance measurements. (Consult the instruction manual.) Set the function selector switch to the ohms position. Select the proper range with the range switch. The ranges are usually marked as multiples of R. For example,

$$R \times 1 \quad R \times 10 \quad R \times 100 \quad R \times 1 \text{ k}\Omega$$

The value of the resistor can be found by multiplying the reading by the range setting. For example, a reading of 11 on the $R \times 1$ kΩ range is 11×1 k$\Omega = 11$ kΩ, or $11,000 \, \Omega$. Note that the ohms scale (usually the topmost scale) reads from right to left, opposite to the other scales on the meter face.

Before you attach the unknown resistance to the test leads, short the leads together and observe the reading. If the meter does not read zero ohms, adjust the ZERO ADJUST control until it does. Disconnect the leads. The meter should now read infinite resistance (the extreme left-hand index of the scale). If the meter does not read infinite ohms, an adjustment will be required, using the OHMS ADJUST control. On the VOM, the control is a mechanical adjustment which requires the use of a screwdriver. Do not attempt this adjustment on the VOM without first consulting your instructor. The zero and infinite ohm adjustments must be checked each time the range of the ohmmeter is changed. Otherwise, the readings will be incorrect.

Proper Use of the VOM Multimeter

1. Always start with the highest range of the instrument and switch down to the proper range successively.
2. Use the range in which the deflection falls in the upper half of the meter scale.
3. Whenever possible, choose a voltmeter range such that the total voltmeter resistance (ohms/volt × FSD) is at least 100 times the resistance of the circuit under test. This rule will prevent erroneous readings due to the loading of the circuit under test.
4. Use an ohmmeter range so that the deflection falls in the uncrowded portion of the scale.
5. Exercise extreme caution when measuring voltages and currents in high-voltage circuits. A safe procedure is to shut down the power, connect the meter, turn the power on, read the meter, and again shut down to disconnect.
6. Try to ascertain the polarity of dc voltages before making the measurement.
7. Never measure resistances in a live circuit. Always shut down before making resistance measurements.
8. Whenever measuring the resistance of a resistor in a circuit, note whether there are any other resistive elements that could cause an error in the reading. It may be necessary to disconnect one side of the resistor before measuring.
9. Check the zero and ohms adjustments each time the range is changed. For proper procedure, see Appendix III.
10. When making measurements, grip the test prods by the handles as close to the lead end as possible. *Do not allow the fingers to touch the prod tips while measuring.*
11. When the instrument is not being used, do not leave it in the ohmmeter function. If the instrument has an OFF position, use it; otherwise, switch to its highest dc voltage range.
12. *Keep the instruments away from the edge of the workbench and away from heat and dangerous fumes.*

V

Scientific Notation and Trigonometric Identities

Scientific Notation

The positive exponent of 10 is equal to one less than the number of places to the left of the decimal point:	The negative exponent of 10 is equal to the number of places to the right of the decimal point:
$10^0 = 1.0$ $10^1 = 10.0$ $10^2 = 100.0$ $10^3 = 1000.0$ $10^4 = 10000.0$ $10^5 = 100000.0$ etc.	$10^0 = 1.0$ $10^{-1} = 0.1$ $10^{-2} = 0.01$ $10^{-3} = 0.001$ $10^{-4} = 0.0001$ $10^{-5} = 0.00001$ etc.

Dimensional Prefixes

Prefix	Multiplier
tera	10^{12}
giga	10^{9}
mega	10^{6}
kilo	10^{3}
hecto	10^{2}
deka	10^{1}
deci	10^{-1}
centi	10^{-2}
milli	10^{-3}
micro	10^{-6}
nano	10^{-9}
pico	10^{-12}

Trigonometric Identities

$$\sin(-\theta) = -\sin\theta$$
$$\cos(-\theta) = \cos\theta$$
$$\tan(-\theta) = -\tan\theta$$
$$\sin(\theta + 90°) = \cos\theta$$
$$\cos(\theta - 90°) = \sin\theta$$
$$\sin^2\theta + \cos^2\theta = 1$$
$$\cos(A + B) = \cos A \cos B - \sin A \sin B$$
$$\cos(A - B) = \cos A \cos B + \sin A \sin B$$
$$\sin(A + B) = \sin A \cos B + \cos A \sin B$$
$$\sin(A - B) = \sin A \cos B - \cos A \sin B$$